SOCIAL THEORY AND THE GLOBAL ENVIRONMENT

Issues such as ozone depletion, global warming, acid deposition, deforestation and species-loss have until recently been primarily matters for natural scientific determination and 'expert' policy prescription. If the social sciences had any role, it was to investigate social impacts of processes and policy responses formulated elsewhere.

Social Theory and the Global Environment emphasizes the ways in which cultural, economic and political values are *already* involved in shaping the definitions of 'environmental problems' for scientific analysis. Providing a much-needed perspective on the relationship between social theory and sustainability, the book examines the challenge which environmental problems and human concerns with the environment represent for Sociology and stresses the necessity for the voice of social science to be heard on the most pressing issue of our time: environmental degradation and its human cost.

The contributors, all international social scientists with long-standing interests in environmental issues, explore the current policy agenda from a critical perspective, emphasizing the need for qualitative studies of the way societies represent the 'environment' in public discourse. In their various ways they call into question purely technical versions of environmental management, and draw upon social theory to construct a broader view of humanity's relation to 'nature'.

Social Theory and the Global Environment is the first volume in the *Global Environmental Change* series published in association with the ESRC Global Environmental Change Programme and edited by Michael Redclift, Martin Parry, Timothy O'Riordan, Robin Grove-White and Brian Robson.

GLOBAL ENVIRONMENTAL CHANGE SERIES
Edited by Michael Redclift, Wye College, University of
London, Martin Parry, University of Oxford,
Timothy O'Riordan, University of East Anglia,
Robin Grove-White, University of Lancaster and
Brian Robson, University of Manchester.

The *Global Environmental Change Series*, published in association
with the ESRC Global Environmental Change Programme, emphasizes
the way that human aspirations, choices and everyday behaviour
influence changes in the global environment. In the aftermath of
UNCED and Agenda 21, this series helps crystallize the contribution
of social science thinking to global change and explores the impact
of global changes on the development of social sciences.

SOCIAL THEORY AND THE GLOBAL ENVIRONMENT

Edited by
Michael Redclift and Ted Benton

Global Environmental Change Programme

London and New York

First published in 1994
by Routledge
11 New Fetter Lane, London EC4P 4EE

Simultaneously published in the USA and Canada
by Routledge
29 West 35th Street, New York, NY 10001

The index was prepared by Simon Pringle.

Phototypeset in 10pt Garamond by Mews Photosetting, Beckenham, Kent
Printed and bound in Great Britain by
Mackays of Chatham PLC, Kent

British Library Cataloguing in Publication Data
A catalogue record for this book is available from the British Library.

Library of Congress Cataloging in Publication Data
Social theory and the global environment / edited by Michael Redclift
and Ted Benton.
p. cm. – (Global environmental change series)
Includes bibliographical references and index.
1. Social ecology. I. Redclift, M.R. II. Benton, Ted
III. Series.
HM206.S57 1994
304.2–dc20   93-44072
                                                      CIP

ISBN 0-415-11169-2   0-415-11170-6 (pbk)

CONTENTS

FIGURES

CONTRIBUTORS

Barbara Adam is a sociologist, working at University College, Wales. She is also currently a Research Fellow under the GEC Programme.

Ted Benton is Professor of Sociology at the University of Essex.

Frederick Buttel is Professor of Rural Sociology, University of Wisconsin.

Cecile Jackson is at the School of Development Studies, University of East Anglia, and is also a Research Fellow under the GEC Programme.

Michael Jacobs is a consultant and environmental economist currently holding a Research Fellowship at CSEC Lancaster University under the GEC Programme.

Angela Liberatore, from Italy, is a consultant with the European Commission (DG.XII). The chapter was written while she was at the European University Institute.

Michael Redclift is a Professor in the Environment Section, Wye College and Research Co-ordinator, *Global Environmental Change* (ESRC).

Elizabeth Shove, a grant-holder under the GEC Programme, is a Senior Lecturer in Sociology, University of Sunderland.

Leslie Sklair is in the Department of Sociology at the London School of Economics.

Peter Taylor is at the University of Cornell.

Graham Woodgate is Lecturer in Environmental Sociology at Wye College.

Brian Wynne is Professor and Research Director, Centre for the Study of Environmental Change (CSEC), University of Lancaster.

Steven Yearley is Professor of Sociology at the University of Ulster, Coleraine.

1

INTRODUCTION

Ted Benton and Michael Redclift

This collection of essays was stimulated by a reawakening of public anxiety about environmental dangers. The global issues of climate change, ozone depletion, deforestation and world food crisis orchestrated this anxiety. But for many individual citizens environmental concern was consolidated through personal experiences of air pollution, congestion and powerlessness in the face of unwelcome 'development', climatic anomalies and dietary worries. All appear at times to be global symbols. The gap between 'lay' perceptions of environmental crisis and official scientific, technical and policy discourses is, of course, one area for sociological exploration. Another quite commonly recognized area concerns the social, cultural and economic *impacts* of ecological change: what will be the effects of climate change on agricultural activity and on the economic status or political stability of the areas affected? What balance of costs and risks are incurred by alternative responses to the likelihood of lowland flooding?

However, the current phase of environmental concern has spawned among social scientists a much wider, more diverse, and more imaginative role for the social sciences in environmental debate. Above all, there is a commitment to exploring the ways in which patterns of social relationships, cultural forms, political practices, and economic institutions are all implanted in the *production* of environmental change. In particular this means, as Howard Newby (1991) has emphasized, challenging the 'technological determinism' which has dominated so much of both the environmental debate and environmental policy formation. A day conference called under the initiative of the Economic and Social Research Council's Global Environmental Change Research Programme in March 1992 was intended as one step, among many others, towards opening out a new diversity of social scientific questions and projects in relation to the environment. Many of the chapters in this book are developed versions of presentations given on that day. Other chapters were commissioned by the editors after the meeting took place.

1

The emphasis is on diversity, imagination, and exploration. There is certainly no common 'line' presented here, although all the contributors are committed to the view that the social sciences have a more significant role to play in understanding and responding to environmental crisis than has been widely assumed in the past. But there is, equally, an acknowledgement – in some cases, an insistence – that the social sciences are not equipped to play this enlarged imaginative and practical role without a radical re-think of their *own* inherited assumptions. There is a two-way process. Environmental debate stands to benefit greatly from the insights of the social sciences, but, equally, the social sciences themselves have much to learn from their attempt to rise to this challenge. There is, we hope, much food for thought in these essays for social and political theorists, sociologists, economists and others *irrespective* of their prior commitment to the 'environment' as a research specialization.

There are two main reasons for this. The first is that serious attempts to come to terms with the issues posed by our environmental crisis expose to critical examination some very basic 'settled' assumptions of the 'mainstream' traditions of the social sciences. Calling into question these settled assumptions may well open up, or reopen, a research agenda for the social sciences which reaches well beyond the specifically environmental issues posed here. The second reason why these essays may be interesting to social scientists who do not see themselves as primarily concerned with environmental issues, is that these issues reflect several long-standing and unresolved disputes within social theory – between agency-centred and structuralist approaches, individualist and holist, and others. Whilst we certainly do not contend that these issues are *resolved* by the confrontation with environmental questions, there is no doubt that the approaches outlined in this collection add new dimensions and new ways of thinking about these troublesome questions.

The collection is by no means preoccupied with these second-order, reflective questions concerning disciplinary assumption and boundaries. We have tried to secure an appropriate balance between critical self-reflection, on the one hand, and direct demonstration of the substantive insights which can already come from the application of social scientific methods and concepts to environmental issues, on the other. The sociology of social movements, policy-analysis, 'global systems' sociology, sociology of science, 'structuration' theory and phenomenology are all put to constructive use in these wide-ranging essays.

A QUESTIONABLE HERITAGE: SOCIAL THEORY AND THE ENVIRONMENT

The prevailing approaches in social and political theory emerged and consolidated themselves in the decades around the turn of the present century. The all-pervasive influence of biological thinking at that time was countered, in the humanist traditions of social thought, by an insistence on human distinctiveness *via-à-vis* the order of nature. Culture, meaning, consciousness, and intentional agency differentiated the human from the animal, and effectively stemmed the ambitions of biological explanation. For Durkheim (1982), the social constituted a *sui generis* reality which interrupted the chain of causality linking both external physical *and* inner biological pressures on the individual. In one move, the opposition between nature and culture (or society) made room for social sciences as autonomous disciplines distinct from the natural sciences, and undercut what were widely seen as the unacceptable moral and political implications of biological determinism. The firm categorial opposition between nature and culture has, of course, been subsequently reinforced by the twentieth century's continuing experience of biological determinism in practice – in eugenic projects of 'racial hygiene', in successive waves of resistance to women's social and political demands, and, most infamously, in the rise of European Nazism (see Benton 1991).

Economics is a social science which has, in many respects, escaped these pressures. Many of the most basic assumptions of the dominant traditions of economic thinking are inherited from the early nineteenth and even eighteenth centuries. Moreover, the preoccupation of economics with wealth-creation, efficiency in production, and the satisfaction of human wants suggests the inescapability of a confrontation with the material conditions and setting of economic activity. It is, perhaps, surprising, therefore, that the concepts and assumptions of 'mainstream' economics have also been inimical to environmental concerns. Many of the nature-given conditions of economic activity have been regarded as 'free goods' which therefore played no role in the determination of prices or in the fluctuations of demand. Even for those goods whose scarcity could not be ignored, the economic *significance* of scarcity has been assumed, in general, to be adequately grasped through concepts such as the economic cost of acquisition. In general, economics has been concerned with quantitative questions about the values or prices of commodities in exchange, at the expense of qualitative questions about what *kinds* of things get produced, *how* they are produced, and with what social and ecological consequences.

So, for the social sciences to approach environmental problems is to overcome some deep-seated (and, in important respects, *well-founded*) inhibitions. If it is once allowed that humans may be considered sufficiently like other living organisms for us to apply to them the concepts of ecology, what is to stop us moving on to the other life sciences? Why not explain gender differences in terms of hormones, brain lateralization, or 'parental investment'? Why not explain private property, war and racism as 'natural' consequences of our 'animal' instincts of territorialism, hierarchy and aggression (Rose *et al.* 1984)? Since the dominant traditions in economics assume that actors rationally pursue preferences which are determined *outside* economic life, these questions do not crowd in on them. This may be one reason why economics has taken the lead among the social sciences in turning its attention to environmental problems. However, as Michael Jacobs' contribution to this collection shows, this lead is not an unproblematic one. For the rest of the social sciences the question remains: how do we open up to investigation the relationships between humans and the rest of nature, without letting in the 'Trojan Horse' of biological determinism?

A second 'settled assumption' of our contemporary social sciences also derives from the historical priorities during which the dominant approaches were formed. This is the commitment to whole societies – usually, but not in all cases, nation-states – as a primary unit of analysis. In cultural anthropology, of course, this became a fundamental methodological principle. In political science, the orientation of political action to the national state has been reflected in the dominant research traditions, with transnational political processes being understood primarily in terms of relations between nation-states. The main traditions in sociology, too, where they have dealt with large-scale social processes, have thought of these in terms of their belonging to more-or-less identifiable particular 'societies'. In twentieth-century Marxism, the concept of 'social formation' as a complex articulation of 'levels' of social practice has served to keep the primary focus of analysis on particular societies, despite a countervailing recognition of the globalizing tendencies of capitalism.

Since the 1960s, however, Marxist-influenced 'dependency' theorists showed that 'development', or the lack of it, in Third World countries could not be understood in abstraction from the location of those countries in an increasingly integrated global economy. Since that time attempts have been made to theorize social and cultural processes at a global level. These attempts have not all been stimulated by a concern for environmental issues, but it is fair to say that such theoretical developments are increasingly relevant. Whilst there are great dangers in giving excessive attention to currently high-profile

4

'global' environmental issues it is still very clear that ecological processes do not respect national boundaries. In some cases (as, for example, acid rain resulting from UK power-stations, or the nuclear fall-out from Chernobyl) environmental damage is initiated in one country, but its effects are felt in one or more others. In other cases (such as global processes like ozone depletion and global warming) industrial or other activities which are widespread in geopolitical terms have cumulative effects at the level of the whole ecosphere. In general, where there is a mismatch between the causes of environmental change, its physical distribution, and the boundaries of nation-states there are pressures for political responses which cannot be adequately analysed in terms of the traditional orientation of political science to issue-formation and decision-making within nation-states. Transnational social, cultural and economic processes, too, must be brought into the picture.

SPACE AND TIME: RENEWING THE CHALLENGE TO A GENERALIZING SOCIAL SCIENCE

Implicit in both of these increasingly contested assumptions is a third: that, in so far as the social sciences have thought of themselves as generalizing, abstract sciences, they have tended to develop explanatory strategies which abstracted from the particularities and contingencies of space and time. This is, indeed, one way in which the social sciences distinguished themselves from the narrative character of history. Of course, both time and place *have* featured significantly in sociology and the other social sciences. Max Weber's investigation of the major world religions was in part concerned with understanding the specifically cultural conditions which enabled capitalist economic development to occur in the West, but not in the East (Marshall 1982, Schroeder 1992). But, significantly, Weber set out to explain where and when economic development took place. The concepts he used to explain these spatially and temporally located processes were the general features of the different religions: their doctrinal contents, modes of organization and relations to other social and political processes. There is nothing about the location of these cultural forms in the Orient or Occident *as such* which is of explanatory relevance.

This space–time indifference of the main traditions of social science can be seen as, in part, a residue of the positivist legacy with its view of scientific knowledge as a deductive hierarchy of abstract laws. However, it is also sustained by the anti-positivist, 'interpretivist' or 'hermeneutic' schools of thought. For these traditions, the focus on meaning, subjectivity, and symbolic forms or 'discourses', tends to abstract these social processes from their physical embodiments, and, consequently from their location in time and space. Again, it could be argued that Marxism, with its concern, in the theory of imperialism and in subsequent work on development and dependency, with the geographical spread of capitalist economic, political and military forms, anticipated the later 'spatialization' of social science. One reason for this may have been Marxism's status as a form of theoretically reflective historiography, rather than as an 'abstract science' in the positivist sense.

However, even within Marxism the integration of spatial and temporal considerations into theorizing has been largely *ad hoc* and inconsistent. A more sustained and self-conscious concern with space as a sociologically relevant variable came in the 1970s with a significant convergence between, on the one hand, a radical human geography and, on the other, a revitalization of urban and rural sociology (see, for example, Harvey 1973, especially Ch. 1). As geographers came to acknowledge the significance of social and political processes, sociologists became more aware of the significance of the fact that social activities had a definite, analysable pattern of distribution in physical space.

Even then, the equally significant fact that social processes have duration through time continued to be given scant attention. As with space, time was always in some sense present in social explanation and theorizing. Much of classical sociology has been either implicitly or explicitly an exercise in periodization – in defining the distinctive features of modernity and exploring their human and moral significance: witness the current intensity of sociocultural debate between advocates and opponents of the thesis of post-modernization. Again, much 'micro' sociological work on the 'life-cycle', employing biography as a method of analysis (see, for example, *Sociology*, special issue, vol. 27, no. 1, February 1993), cannot avoid acknowledging temporal duration as a significant dimension of the social world. Meanwhile, the structural Marxists of the 1960s and 1970s attempted to do away with economic determinism by emphasizing the differential 'times' of the various social practices comprising the social formation. The social theorist Anthony Giddens is notable for his explicit focus on both spatial and temporal aspects of social processes (Giddens 1979, especially 'Introduction'; and 1981, Ch. 1), whilst a major step forward was taken by one of the contributors to this volume, Barbara Adam, in her path-breaking *Time and Social Theory* (1990).

Though it is arguable that recent attempts to integrate both time and space more fully into sociological analysis were not primarily motivated by environmental issues, there is no doubt that these developments are an absolutely basic necessity for any sociology to rise to the environmental challenge. As we have already noticed, the spatial distribution of environmentally relevant human social action, as well as of its ecological *effects*, bears no necessary relationship to the boundaries within which legal and political decision-making or authority operate. So, being able to integrate a spatial dimension into social scientific analysis is essential if these dislocations are to be analysed. Similarly, this applies to temporal duration. The spatial diffusion of air and water pollutants; the passage of chemical waste products into, and then through, food chains; the processes of radioactive decay; the advance of scientific research and technical innovation; processes of cultural change, democratic decision-making, international negotiation, legislation and regulation are all processes which take time. But the time-scales on which they operate are indefinitely variable and, again, bear no necessary connection with one another. The complexities and dislocations of these interacting time-scales would somehow have to be brought within social scientific analysis for any adequate understanding to be achieved. The implications for any environmental politics, too, are clearly immense. Finally, it should be noted that full integration of spatial and temporal aspects of social and ecological processes into social scientific analysis is actually a necessary condition for social science to go beyond its inherited nature/society dualism, and also for it to move beyond its confinement to the nation-state as its primary unit of analysis.

STRUCTURE AND AGENCY IN THE DEVELOPMENT OF ENVIRONMENTAL CONSCIOUSNESS

So much, then, for those aspects of the shared conceptual inheritance of the social sciences which must be called into question if environmental issues are to be adequately addressed. There are also several areas of long-standing and quite often acrimonious dispute in social and political theory. In some respects these issues may be rendered more tractable in the effort to engage with the environment. Even where no such benefit accrues, the relationship between social processes and their environmental conditions and consequences may at least serve to shift the terms of debate and offer new understanding of the significance of these 'old' conflicts.

The first and most pervasive of these contests is that between approaches which put human conscious agency at the centre of analysis, and those which focus attention on the social-structural conditions for, and constraints on, action. One of the most important insights which the social scientist can offer in the environmental debate is that the eminently rational appeals on the part of environmentalists for 'us' to change our attitudes, or lifestyles, so as to advance a general 'human interest' are liable to be ineffective. This is not because (or *primarily* because) 'we' are irrational, but because the *power* to make a significant difference, one way or the other, to global, or even local environmental change is immensely unevenly distributed. In the face of this, the great majority of individuals may be 'locked into' patterns of daily activity (such as private car use) which they know to be environmentally destructive. Only if the spatial separations of work, shopping, leisure and residence were changed, or if public investment into socialized transport provision were greatly increased and transformed would individuals have a meaningful *choice* to make about transport options for themselves. This structuralist emphasis on the patterned contexts in which individuals make choices clearly has a lot to offer for environmental analysis. However, it is equally clearly true that environmental social movements have, at least in some countries, affected public cultures and authoritative decision-making in ways which show that a thoroughgoing structural *determinism* is mistaken. Under what conditions can individual or collective agency make a difference? Can forms of collective decision-making be developed which have some prospect of transforming social structures in an environmentally benign direction?

LEVELS OF EXPLANATION IN THE CONSTRUCTION OF THE 'ENVIRONMENT'

Closely related to the 'structure/agency' dispute is that between advocates of holist and individualist approaches to explanation in the social sciences. Individualists generally claim that 'society' is nothing over and above the individual people of which it is composed. To the extent that it is appropriate to speak of social phenomena at all, we should think of these as *aggregates* of individuals and their activities. In sociology, Max Weber is often used as an exemplar of this approach although he was certainly not a *consistent* individualist. Economics, especially in its currently dominant 'neo-classical' form, is pre-eminently individualist in its approach. Individualist approaches are most often countered in sociology by 'social realist' arguments deriving from Durkheim or from Marx, whilst the 'whole culture' approach which predominates in anthropology is a bulwark against individualism in that discipline.

These opposed strategies for social scientific explanation clearly have profound implications for both environmental research and policy. Individualists will tend to focus on individual demands, desires, or decisions as the ultimate basis of environmental changes, and so will tend to recommend policy options which give individuals incentives to act in environmentally benign ways. Holist, or 'social realist', approaches will tend to emphasize ways in which individual behaviour is shaped by the wider collectivities or normative frameworks within which individuals are situated. This approach would emphasize the weaknesses of 'neo-Malthusian' forms of environmental analysis which estimate and predict human environmental impacts by simply aggregating and extrapolating individual-level impacts (whether in terms of population growth, resource-utilization or pollution-impact). One important way in which environmental issues affect the terms of this debate is by showing that societies are not just composed of individual human beings. It is necessary to analyse and explain the ways in which human social structures bind together not only people but also (non-human) animals, physical objects, spatial 'envelopes', and so on.

A third area of dispute amongst social scientists is much less easy to characterize. In some ways, it is closely related to the above two disputes. An imaginary 'extreme' structuralist might argue that the social actions of individual people are predictable on the basis of our analysis of the structural conditions under which they act. On such a view, actors' *conceptions* of their situations, and their creative capacity to devise new goals for themselves, or think up new ways of achieving what they want, would play little if any part in social explanation. By contrast, within actor- or agency-oriented approaches, these features of knowledgeable and creative agency would be seen as the primary materials for social explanation. In the environmental debate, there are loose parallels between these opposed approaches and the opposition between 'technocratic' perspectives and ones which emphasize the role of popular culture and 'lay' knowledge. In assessing the risks attaching to new technologies, for example, it may be argued that there is a scientific basis for assigning a definite numerical value to 'objective' risk, and developing public policy as 'rational' as a consequence. However, as Mary Douglas (1985, 1990) and others have argued, the cultural setting in which this takes place will necessarily be one in which people assign definite symbolic significances or values to harms of different kinds. These will yield action-preferences which may be systematically at odds with official ones, and be deemed, from that standpoint, to be 'irrational'.

The environmental debate itself, in its contemporary forms, rests upon differential 'take-up' of scientific knowledge claims by policy communities, publics of various kinds, social movement activists, party leaderships and so on. What are the processes of communication, discursive 'processing', normative orientation, 'moral entrepreneurship' by which these public antagonisms get formed and transformed? These are questions which place the environmental debate itself into contact with social scientific research traditions in the sociology

9

of science, sociology of culture, and sociology of social movements, with potentially illuminating consequences for each. Not only will 'official' science get no easy endorsement from the sociology of science, but neither will the knowledge-claims of the environmental social movements. This may be disconcerting to both. Equally, however, at the risk of abolishing their subject matter altogether, sociologists of science and of environmental social movements will have to abandon their carefully nurtured stance of cognitive relativism or agnosticism. Is it possible to take a non-committal stance on the status of the scientific knowledge claim that CFCs are a cause of ozone depletion, whilst sociologically evaluating the political response to it? Undoubtedly, this is an area which will provide a fruitful testing-ground for rival relativist and realist approaches to the sociology of science.

POST-MODERNISM AND THE ENVIRONMENT

Finally, there has in recent years grown up an intense controversy in the social sciences about how to characterize the present condition of social life. In part this has arisen from a turning away from concern with large-scale social and historical processes in favour of a micro-sociology of subjective life, and associated linguistic and cultural processes. Through this, the social sciences have become open to methods of analysis previously developed in psychoanalysis, structural linguistics, and cultural criticism. A widely shared sense of deep-level shifts in self-identity, language use and aesthetic styles, which already pervaded those disciplinary fields, then began to enter more specifically social scientific debate.

Could the break with aesthetic modernism, the interpenetration of 'high' and popular culture, the sense of disorientation, and inability to assign overall meaning to social life and its direction, including the fragmentation of self-identity, be defined as the condition of 'post-modernity'? Could it be correlated with economic, social or political changes? Are human societies now passing through a transition to new forms in which the inherited social scientific categories and methods of explanation are no longer any use? Alternatively, are these widely proclaimed 'new' experiences merely intensifications or 'accentuations' of the fragmentation and invasion of subjective life typical of earlier phases of capitalist modernity, and theorized in the past as 'estrangement', 'anomie' or 'reification'?

This debate has important implications for our concern with the relationship between society and the environment. The post-modernism thesis is often associated with 'post-industrialism': the claim that, increasingly, economic activity is devoted to the provision of services, as against industrial production, and that the new communications and information technologies entail quite new relationships to work, and patterns of culture and consumption. If this view is right, it suggests that the projection of past rates of industrial growth, resource-use and pollution into the future may not be justified. Alarming predictions of environmental collapse may have been based on a failure to recognize that historical processes are themselves taking us on to a more environmentally benign future. Set against this line of thinking, however, is the emphasis (expressed in several contributions to this collection) on *global* social and economic processes. At least one of the advocates of post-modernism (Baudrillard 1989) concedes that only one society, the USA, is yet fully 'post-modern'. We need to ask whether the global environment will bear the costs of the rest of the world's economies in following the US example. More seriously still, we need to investigate how far the higher environmental standards and shifts towards service economies in some of the most 'developed' countries of the North themselves depend on the export of pollution and labour-exploitation to the poorer countries and on the continued flows of capital and raw materials from poorer to richer countries. Finally, the tendency amongst post-modernists away from 'out-dated' political economy, and their rejection of 'truth-seeking' as the purpose of intellectual work, would rule out just these very necessary investigations.

DISCIPLINARY BOUNDARIES AND THE ENVIRONMENT

So far, this introduction has emphasized ways in which the legacy of social scientific assumptions may have to be called into question if environmental issues are to be adequately addressed. The bearing of environmental issues on long-standing unresolved disputes within the social sciences has also been briefly addressed. But we have also tried to strike a balance in the essays collected together here between these critical reflections on the social sciences themselves and the more positive and urgent task of putting social scientific ideas and methods to work in the understanding of our environmental crisis.

Many of the essays in this collection are either avowedly inter-disciplinary or they explicitly take issue with the assumptions of individual social science disciplines. It might be useful to put this unease with conventional disciplinary boundaries into an intellectual context.

As Figure 1.1 suggests, research on the environment occurs at different points of convergence between disciplines and with the evolving field of environmental studies. Within the physical sciences

11

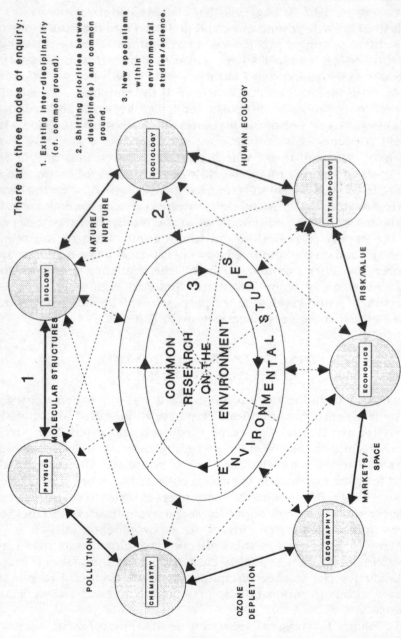

RESEARCH ON THE ENVIRONMENT OCCURS AT DIFFERENT POINTS OF CONVERGENCE

There are three modes of enquiry:

1. Existing inter-disciplinarity (of. common ground).

2. Shifting priorities between discipline(s) and common ground.

3. New specialisms within environmental studies/science.

SOCIOLOGY

ANTHROPOLOGY

HUMAN ECOLOGY

BIOLOGY

NATURE/ NURTURE

2

COMMON RESEARCH ON THE ENVIRONMENT

3

ENVIRONMENTAL STUDIES

RISK/VALUE

PHYSICS

MOLECULAR STRUCTURES

1

ECONOMICS

POLLUTION

MARKETS/ SPACE

CHEMISTRY

OZONE DEPLETION

GEOGRAPHY

Figure 1.1 Environmental research

environmental 'problems' frequently transcend disciplines. Issues such as pollution, the depletion of the ozone layer, or changes in molecular structures often require investigative approaches from several science disciplines. (In Figure 1.1 this is located between Physics and Chemistry, for example.) These 'science' issues have a call upon the social sciences as well. First the acknowledgement that environmental problems cannot be 'solved' by the natural sciences has brought some of the social sciences closer to a 'problem-centred' view of the issues. Economists, particularly, have sought to develop their expertise in relation to a growing body of research in environmental studies and environmental science. The 'management' of the environment assumes urgency as we become more aware of what is going wrong in our relationship with the natural environment.

At the same time the discourses surrounding the environment have prompted shifts within the social sciences that are linked to the 'science' agenda, but which are frequently important in their own right. For example, following Figure 1.1, problems of human ecology have been located between anthropology and sociology. However, a number of disciplines, including anthropology, sociology and geography, have each arrived at points of convergence from different points of departure. In similar fashion, environmental concerns, such as risk analysis or the competing claims of 'nature' and 'nurture', continue to evolve in the interdisciplinary interstices (between economics and anthropology in Figure 1.1) that characterize environmental debate.

Finally, a growing body of knowledge and explanation is to be found within intellectual territory that can only be described as 'common ground'. The interpretation of the environment in the social sciences assumes territoriality of its own; reflected in new 'environmental' journals, educational qualifications and professional titles. Although the essays in this book were not written as an exploration of the breakdown of disciplinary boundaries they do reflect the fact that this process is well advanced. Asked to identify the discipline from which social theorists of the environment originate one would need to move beyond sociology itself to geography and anthropology and, as this volume illustrates, economics and international relations.

GLOBAL ENVIRONMENTAL CHANGE
AND THE SOCIAL SCIENCES

The discussion of global environmental change has been dominated by the physical sciences, with consequences that are often problematic for the social sciences. The deliberations accompanying the work of the Intergovernmental Panel on Climate Change (IPCC) are a case in point. The IPCC established three Working Groups to report on the evidence for global warming: Scientific Processes, Impacts and Responses. Even nominal representation of social science was never achieved within these groups. Even more alarming, however, was the linear model of cause and effect which underlay the organization of the scientific research. It was confidently expected that the discoveries of science could lead to an analysis of their impacts, and policy responses. The role of the social sciences, in this ordering, was exclusively to devise policy responses. The principal claim of the social sciences – that they could explain and interpret human behaviour towards the environment – was left to one side. The anthropogenic *causes* of global warming, exemplified in every aspect of our 'getting and spending' and around which the social sciences had evolved, were tacitly ignored in the science-driven approach adopted by IPCC.

A contrast with the Report of the Brundtland Commission, *Our Common Future* published in 1987, makes the point clearly. Where the IPCC reports sounded the warning about what might happen to global climate, the Brundtland Commission laid emphasis on the human activities, already unsustainable, which drove the engines of global warming. In addition, by prompting more discussion of 'sustainable development', Brundtland sought to enlist the natural sciences in support of human purposes and designs. It was clear that sustainable development conformed to human needs and values: it was understood differently in different cultures, and by different interests and groups within societies (Redclift 1993). The direction taken by science, and the uncertainties surrounding scientific 'knowledge', were closely paralleled in the uncertainty experienced by human beings faced with inadequate knowledge of their own *immediate* environment. If dealing with uncertainty distinguished 'everyday' environmental choices, how much more important it has proved to be in determining *global* changes.

Much of the early discussion of global environmental issues failed to consider the cultural dimensions of global change. This leads us to consider globalization as a *cultural* process linked to changes in the environment, but distinguishable in its own terms. The global reach of environmental problems is paralleled by important changes in the way the environment is understood (King 1991, McCormick 1989). The term 'globalization' implies interconnectedness, but it is clear that images and representations flow in different directions. There are different dimensions of global cultural relations, with specific, and interconnected, implications for the environment. Of the four dimensions which come to mind – the spatial, technological, material and representational – each can be considered as a component of globalization.

The evolution of a global car industry, for example, is clearly linked with spatial factors and material resources, but the cultural process of globalization extends to the way automobiles are marketed and represented as cultural 'icons'. The environmental consequences of globalization are linked to all these factors. On the one hand it has been suggested that spatial relocation of the Japanese car industry throughout the globe has improved air pollution in Tokyo. At the same time, opening up new consumer markets for Japanese cars, and meeting global consumer 'needs', implies increased per capita consumption of energy at the global level. Technological diffusion presents new problems of waste disposal, and makes new demands on the conservation of resources. Economic globalization redistributes environmental costs and benefits.

It is also clear that there is an ordering of the global agenda, which is dominated by the industrial North (Redclift 1992). The way that economic development is conventionally understood suggests that the North prefigures the South as a market for goods and services: it is only a matter of time before access to the goods enjoyed by consumers in the North is widened to include many in the South. And yet this view of development is increasingly inadequate. The 'limits to growth' in the 1970s were judged to be materials shortages and food short-falls. From the perspective of the 1990s the view is different: today it is the 'externalities' of growth (particularly air pollution and water quality) which provide the 'limits' (Meadows *et al.* 1992). Patterns of consumption are driven by effective demand in the North; but the outcome of increased consumption of resources, in the form of changes in climate and loss of biodiversity, is felt throughout the globe. In addition, there is wide, and increasing, divergence between the Northern 'environmental' agenda, and the 'development' agenda shared by most poor countries. The meeting of the United Nations Conference on Environment and Development (UNCED), in Rio de Janeiro in July 1992, only served to underline these differences.

The world's population is expected to double between 1985 and 2025: from 4.8 billion to 8.2 billion in the space of forty years. As the second World Conservation Strategy (1991) made clear, one quarter of the world's population, in the North, consumes 80 per cent of the commercial energy produced. In the South, three-quarters of the world's population consume just 20 per cent of the commercial energy. The per capita consumption of commercial energy in the United States, in 1991, was ten times that in China and 140 times that in Bangladesh (*Caring for the Earth* 1991). Increases in aggregate population, concentrated in the South, will increase the demand for energy to provide for basic needs. In the North, where energy policy is 'supply led' and we are encouraged to produce as well as consume

15

more energy, there are enormous potential savings in energy production (IIED 1991). As the Intergovernmental Panel on Climate Change
(IPCC) argued, making energy production more efficient in the North
requires a legal and policy framework which is presently lacking.

Environmental awareness in the North, however ambiguous its
origins, marks a shift in thinking about development. Increasingly,
policy interventions seek to limit the environmental damage associated
with industrial growth, and the risks associated with technological
change. There are broadly two views of these processes. Increased
sustainability must be incorporated within products and services
themselves, through better environmental management and techniques
such as Life Cycle Analysis (LCA), or they will represent a threat which
can only be averted by abandoning our current technological imperatives, and adopting radically different and much 'greener' forms of
social life. This includes 'life-style' change, but also could include more
radical views of social structural change. The discourse surrounding
sustainability in the North is caught uncomfortably between both
approaches. The first discourse is that of scientists and governments;
the second that of the Green Lobby and, to some extent, the public.

Development concerns are also a driving force in the South, but
in a way that could hardly be more different. Increases in per capita
consumption of energy and resources are needed to bring most of the
world's population within reach of basic living standards, but meeting
these needs seems unlikely without compromising the needs of future
generations. By 1990 some 60 per cent of the world's population had
access to a television set. Through television the way 'development'
is presented is more global than ever before. In the villages of India
and Peru people watch *Dallas* on television. It is difficult to separate
such media messages from the environmental implications, particularly
in a world where consumer tastes are fashioned for global markets.
Today's urban migrant is often a television viewer, as the television
aerials adorning the shanty towns of Caracas and Mexico City bear
witness.

These examples should prompt us to consider how the experience
of the global environment, conjured up through the media and mass
communications, compares with the reality of the environment with
which most poor people are familiar. Do we stand 'inside' our concept
of the global or 'outside' it? Do people in other cultures? Does
globalization submerge difference or simply highlight it? Do we share
an understanding of the global environment in the same way as we
'share' the globe? Most of the writers contributing to this volume have
sought to deconstruct the 'realities' of the environment, placing
emphasis on the way in which its representation serves to bolster
the influence of various stakeholders: scientists, politicians, interest

groups and the public at large. We should not forget, however, that most of the world's population lies outside, or at the margins of, this discourse.

There is an alternative model of global environmental change from that of bodies such as IPCC. In this model 'science' is linked to 'culture' and 'knowledge' to 'policy'. In other words there is no scientific agenda that is not *also* a cultural artefact, and no way of viewing policy which divorces it from its roots in societies and cultures. Taking the ecosphere as a whole, at the highest possible level of geographical resolution, we need, nevertheless, to identify what underpins global change. The 'problems' in the environment, and their anticipated 'solution', depend on the level at which environmental processes are specified: at locality, regional, national or international level. This level of problem specification, in turn, provides divergent lines of intellectual enquiry. There are various 'epistemic communities' which correspond, as we have seen, with the 'environment' in the social sciences. For an anthropologist the 'global' might involve the world view of an individual; for an international lawyer the relations between nation-states.

In addition, problems of scale are *temporal* as well as spatial. We need to examine human behaviour over time. As Barbara Adam reminds us, this implies re-examination of many of the assumptions that determine our understanding of everyday life. Societies vary in the weight they attach to future generations, and social science disciplines attach different weight to future generations in their intellectual armoury.

Figure 1.1 also points to the fact that although global processes of change might produce more homogeneity they might equally lead to increasing diversity. Even if we confine ourselves to the physical dimensions of global environmental change, such as global warming and biodiversity losses, their effects within societies will help determine the form that changes take. It is human beings, as we have argued, that mediate relations with nature, within specific structural contexts. Culture and knowledge are not merely determined by science, but serve to fashion science and policy themselves. These concerns are evident in most of the chapters in this volume, and represent the focus from which many of them are written.

The structure of the volume reflects a distinction between theoretical criticism and the need to address immediate policy issues. Chapters 1–4 are concerned with the more 'second order' problems: the critical task of investigating the ways in which existing disciplinary boundaries and assumptions need to be questioned. Chapters 5–10 place the emphasis on the positive contribution which the social sciences can already make to the understanding of environmental problems. However, each of these contributions also sounds a note of caution concerning the existing disciplinary legacy.

Ted Benton's contribution (Chapter 2) reviews some common patterns of thinking in the environmental debate. He tries to show that, whilst technocratic *and* many 'deep Green' approaches do not give sufficient attention to social, economic and political aspects, many of the most influential social scientific approaches tend to go to the opposite extreme, barely acknowledging the independent reality of 'nature' and the 'environment' at all. It might seem that a middle way, recognizing the independent reality of both society *and* nature, then studying their interconnection, would resolve this issue. However, Benton argues that this sort of approach tends to be quite difficult to maintain, tending to slide over into an 'over-socialized' position. He argues, more positively, that our social relationships to nature should be thought of in terms of *specific* social practices along with the environmental conditions (physical space, fresh air, raw material, etc.) and media (tools, machines, bodily activity itself, and so on) which are necessary for them to be carried on. This avoids the abstractions of 'society' and 'Nature', and builds into the basis of the approach the mutual dependence of social practices and their non-human conditions. This leads to the suggestion that physical objects and substances, spatial relations, non-human animals and plants may all be theorized as belonging to the social as objects, conditions and media of social activity. At the same time, they are never *wholly* incorporated into society, and persist as complex orders of causality which both enable and constrain human social activity in ways which are only partially calculable and predictable. Limits to growth, in this view, should be understood not as imposed by the finitude of 'external' physical reality but as arising from such unacknowledged material conditions of social practice. In this way, 'limits' can be understood not as more-or-less distant hypothetical dangers but as already pervasive self-induced obstacles and frustrations within the conduct of social life.

Michael Redclift and Graham Woodgate (Chapter 3) begin by noting the neglect of relations between society and the environment in the work of the 'founding figures' of sociology. Nevertheless, they argue strongly that there is much to be learnt from the attempt to apply sociological insights in this field. They use Durkheim and Weber, respectively, as paradigms of the opposed structure- and agency-centred approaches within the discipline. They point out that as, in more recent years, social scientists *have* given more attention to environmental questions, they have continued to do so in a way which maintains the tensions between structure and agency. Human ecologists, urban sociologists and some geographers and anthropologists, have tended towards 'environmental determinism', whilst the interpretative tradition has continued with its view of people as goal-oriented, as 'constructing' views of nature, and interacting with it on the basis of those constructed views. They take a positive view of the concept of 'structuration' developed by the sociologist Anthony Giddens. This concept is intended to incorporate and go beyond the structure/action polarity, and, although not invented to deal with environmental issues, Redclift and Woodgate think it can be readily applied here, too. The environment itself can be thought of as a structure which both constrains and enables action, *and* may be changed by social action.

With this approach as their starting-point, Redclift and Woodgate go on to consider nature/society relationships from two interrelated perspectives. The first, which draws on significant recent work by Richard Norgaard, is concerned with the 'material' aspects of nature/society relationships. Using the concept of 'coevolution' of social life and environmental relationships, they indicate how Western science, technologies, urban life, industrial production and so on have to be seen as a mutually interconnected and reinforcing system which changes through time. The historical development of agriculture and the food system serves as an example of this. But changes in attitudes to, and cultural valuations of, 'nature' also take place alongside, and interwoven with, these 'material' aspects of the coevolution of nature and society. These, too, can be analysed sociologically. Nash's view, that societies come to value nature and wilderness more as their industrial development destroys it, is subjected to some searching questions. In particular, how far can we generalize from Western cultural values and assumptions about nature to other societies, particularly to the economically poor countries of the South? In many of these societies the 'conservationist' ethic imported by colonial powers has left a legacy in which development and environment are seen as contradictory objectives. As Redclift and Woodgate conclude, the concept of 'sustainable development' is an attempt to move beyond this apparent impasse, but whether it will succeed will depend on future research in which sociologists should play a crucial part.

As we have already noticed, economics has taken the lead among the social sciences in addressing the challenge of the environment. A discipline of 'environmental economics' is now well established, and its policy proposals are already influential, both with national governments and international bodies. Michael Jacobs (Chapter 4) provides a sympathetic but very telling critical analysis of the most influential tradition of environmental economics. This approach is based on assumptions it shares with the dominant neoclassical paradigm in economics, and seeks market-oriented solutions to perceived environmental problems. Jacobs eloquently states the key assumptions of the neoclassical paradigm – its claim to value neutrality, its focus on the self-interested rational choices of individual economic actors, and competitive markets which optimize allocations of resources in society. Interestingly, as Jacobs points out, both technology and individual preferences are treated as 'exogenous variables' – as outside the sphere of economic analysis. There is a close parallel here with the technological determinist approaches criticized in Benton's chapter.

Neoclassical environmental economics argues that environmental goods are undervalued, or not valued at all, and so get overused or destroyed. Their suggestion is that if the environment *were* assigned value, which had to be taken account of in real economic decision-making, then this would lead to its being protected at a level which optimized costs and benefits. Since environmental goods tend not to be privately owned, or to be bought and sold in *actual* markets, the neoclassical approach seeks to create hypothetical or artificial markets in them. This, on Jacobs' account, involves three stages – 'breaking down' the environment into particular 'goods', assigning an 'imputed' value to them, and then intervening to achieve the appropriate level of protection by the most cost-effective means.

Jacobs offers some powerful criticisms of each of these stages. Environmental goods are, in general, not at all obviously like actual commodities. They approximate much more to 'public goods' – such as national security or policing – which neoclassical economists already recognize as not susceptible to market-oriented approaches. The methods economists use to estimate 'demand' – such as surveys of public willingness to pay for the protection of some environmental good, or to accept compensation for its loss – are often difficult to interpret. Jacobs argues that high rates of refusal to answer, or 'don't know' responses, as well as discrepancies in responses depending on which questions are asked, show that the respondents' ways of valuing the environment are more complex than a simple monetary valuation. He also offers suggestive evidence that institutional features of business managements and households (for example) affect decision-making in ways that are not predicted by the individual 'rational choice' model of the neoclassical approach. Taking these institutional and sociocultural factors into account, it may well be that, for example, direct regulation might be more effective than market-oriented incentives such as taxation and tradable permits. Moreover, despite its proclaimed value neutrality, the neoclassical approach favours cost/benefit efficiency, whilst abstracting from other possible goals of environmental regulation, such as fair distribution of economic and environmental goods and bads, or recognition of moral rights – including rights of other species. Jacobs concludes with a plea for a more integrated interdisciplinary approach, in which economists would work together with sociologists, political scientists and others in the adoption of a more modest perception of the proper role of economics.

The neoclassical approach to time also comes in for criticism by Jacobs, and this becomes the central theme in Barbara Adam's contribution (Chapter 5). She, like many other contributors to this debate, points out that existing approaches will need to be radically rethought, that pervasive 'dualisms' – of subject/object, mind/matter and so on – will need to be transcended, and that interdisciplinary co-operation will be needed. For Adam there is a danger that prevailing conflicts of views about the environmental crisis will harden into inflexible and polarized oppositions. One fruitful way of side-stepping these, she argues, is to focus on our 'taken-for-granted' assumptions: in this case, unacknowledged temporal aspects of environmental problems.

Adam contrasts the 'embedding' of living organisms and their temporal cycles and 'rhythms' within their ecosystems with human artefacts, whose temporal aspects are 'disembedded' and isolated. This disembedding is intensified with science-based machines. The historical emergence in the West of linear perspective and clock-time, and their consolidation in science-based ways of life, has brought with it our familiar oppositions of subject and object, and our emphasis on control, predictability and proof. This world view, Adam argues, is now being called into question, but is in any case ineffective in the context of the global environmental problems that now have to be faced. One source of hope is that most of us, in our everyday lives, do have ways of thinking and acting which move between different time-scales. It may be that with this as a model a re-engagement with the environment, which breaks through divisions between the personal, the professional and the scientific, may be achieved.

Cecile Jackson's contribution (Chapter 6) tackles the vexed question of gender analysis and the discourses surrounding 'environmentalism'. She criticizes the way in which women have been conceptualized in the South, arguing that Northern environmentalism is largely gender-blind and, in so far as gender identities are recognized, works with reductionist stereotypes as a substitute for gender analysis. The chapter also attempts to indicate the outlines of a gender analytical framework. One consequence of the absence of gender analysis in the environmentalist discourses is the failure to recognize that the environmental relations of women reflect prevailing gender ideologies and struggles. 'Positive' environmental management is frequently based upon coercive social relations. Another consequence of the absence of gender analysis is that the assertion that environmental degradation is caused by 'poverty' remains unchallenged and unqualified.

21

One aspect of the environmental agenda which has already attracted a great deal of attention from sociologists and political scientists is the emergence of environmental social movements themselves. Steven Yearley (Chapter 7) brings together insights achieved by this research in his wide-ranging contribution. He distinguishes between two major research traditions. The first, primarily European, defines social movements in terms of their potential for large-scale social and historical change, whilst the other, mainly US-based, is more 'disinterested' and defines social movements in terms of organizational characteristics. Yearley is critical of both traditions and argues that there is no single criterion which can explain the formation of a large-scale social movement. He prefers to work with a more open-ended descriptive concept which can be applied to research on a case-by-case basis.

Yearley goes on to consider the distinctive features of the environmental social movement. He gives attention to three of these – internationalism, the role of science in the movement, and the presence within it of a vision of a radically alternative society. It is the last characteristic which provides its attraction for the European school of social-movements theory, but Yearley points to the powerful de-radicalizing pressures upon Green politics. The role of science, too, is somewhat ambiguous and paradoxical. The knowledge-claims which give credibility to the Green case are invariably based on scientific research, but it is also true that scientific evidence and standards of proof are also appealed to by the opponents of the Greens. Yearley makes the interesting point that the conduct of the debate on scientific terrain has had important effects on the relationship between professional campaigners and ordinary movement members, and on the style of campaigning activity itself. Finally, Yearley turns to a discussion of the internationalism of the environmental movement, and, in particular, to the difficulties of generalizing from the social movements of the 'developed' countries to the Third World. The great differences in the complex of problems faced in those countries clearly indicates a different pattern of emergence of environmental social movements, and the necessity for more research.

The plurality of cultural values in the face of the global knowledge-claims of environmental science is the theme of Chapter 8 by Brian Wynne. He uses case-studies of rival approaches to global climate change, radiocaesium dispersal and marine pollution, to illustrate the kinds of illumination to be gained from the sociology of science. This discipline is itself characterized by a number of alternative approaches, which Wynne divides into two general orientations. In one sort of approach, which he calls 'interests-based', it can be shown that scientific knowledge-claims and research agenda express either the interests of specific scientific research communities, or wider economic or sociopolitical interests, or, commonly, some combination of the two. Wynne cites the work of Buttel and Taylor as exemplifying this approach when they show the climate models used by the Intergovernmental Panel on Climate Change as reflecting developed countries' interests.

Wynne does not argue against this kind of approach, but he clearly thinks that an alternative 'culturally rooted' approach can take us even further into a 'deconstruction' of established environmental science. Like Buttel and Taylor (see also Benton, Chapter 2, on this), Wynne emphasizes the indeterminacy in scientific knowledge-claims *vis-à-vis* their evidential basis. This goes well beyond orthodox notions of scientific uncertainty according to which this is seen as a temporary problem of insufficient evidence. The kind of indeterminacy that Wynne is concerned with includes questions about how a particular phenomenon is classified as coming under one general definition rather than another; how far causal relationships established under controlled laboratory conditions can be assumed to apply under the many different kinds of contextual conditions which hold outside the laboratory; as well as a whole range of questions about hidden normative and evaluative assumptions *within* science.

The hardening of a scientific consensus around a particular way of constructing an environmental problem such as global warming may appear to result from the evidence but could, in fact, be a result of mutual reinforcement by networked scientific and policy communities. The interest-based approach also has a role to play here, as clearly a consensus of this kind is likely to be more authoritative with public opinion. This is where Wynne's point about cultural diversity comes into its own, however. *How* authoritative a scientific consensus is with any given public, and, still more, how that public understands the relation of that consensus to its *own* disposition to act in various ways, will depend very much on cultural values (including a sense of alienation from 'official' science itself), identities, and traditions. The more global the reach of environmental science's knowledge-claims, the more deeply this question of cultural diversity and the legitimacy of science itself bites. An open-ended, pluralistic, and participatory-democratic context for the formation of a research agenda and policy is implied by this approach to the sociology of environmental science.

This volume does not include any explicit treatment of risk theory, although risk is alluded to by several authors. Our hope is that future writing will take more account of risk assessment, since it underpins the way that science is received by the public, and helps explain widespread inhibitions about changing behaviour and acting along more 'precautionary lines'.

23

Angela Liberatore's contribution (Chapter 9) gives us a valuable case-study of the relationship between scientific research on an environmental issue – global warming – and shifting policy priorities in the EEC. Her study shows that whilst there *are* analysable connections between research and policy, the simplistic model of one-way causal relationships (in either direction) will not do. The study reflects interestingly on the arguments presented by Wynne and by Buttel and Taylor. It is also significant as a study of transnational research and political co-ordination which nevertheless focuses on the regional rather than on the global level.

Liberatore shows that global climate research was being funded by the EC, but was regarded as marginal prior to 1988. When the Commission identified global warming as a political priority it did so on the basis not of its own research but of that of the US and IPCC. There was then a major shift in EC research priorities in favour of basic research on global-warming effects, renewable energy, energy efficiency and earth observation. One very interesting feature of the interrelation between policy and research which Liberatore reveals is very much in line with Wynne's argument. The linkage of scientific research and policy is possible only if problems are constructed by research in a way which renders them definable and amenable to technical or policy interventions of an acceptable kind.

The emergence of a global politics in response to a perception of environmental problems as global in scope sets up a daunting challenge to sociologists. Is it possible to transcend the traditional limitations of even 'macro' sociology to the comparative study of individual societies and to construct a *global* sociology? Is this a coherent enterprise? Do the aspects of social life which sociologists take as their subject matter (normative orders, stratification systems, social divisions of labour, and so on) extend to the global scene? One of the very few attempts to answer this question with a positive theory of global social processes has been the work of Leslie Sklair. In his contribution to this book (Chapter 10), he takes the ambitious further step of interpreting global environmental struggles in the light of that social theory. Sklair's proposals will undoubtedly be very controversial, but they have the great merit of posing bold and clear challenges to empirical research on some central questions of environmental politics.

Sklair opposes the widespread tendency to analyse global environmental processes in terms of inter-state relations, favouring, instead, a conception of 'transnational practices' which may be political, economic or cultural. In each of these domains we can identify dominant institutions – which Sklair calls the transnational capitalist class, the transnational corporation and the culture-ideology of consumerism, respectively. Together these comprise a global capitalist system whose expansionary dynamic is the principal source of threats to the global environment. Corresponding to these dominant institutions of global capitalism, Sklair postulates transnational environmental elites, transnational environmental organizations and the culture-ideology of environmentalism. A good deal of Sklair's chapter is devoted to filling out and explaining these concepts by way of the empirical literature.

Central questions posed by his approach are whether and in which respects there exists a global environmental 'system' which shadows the various elements in the global capitalist system, but as (in some sense) its global opposition. This might be sustained as a plausible view given what seems to be Sklair's fundamental assumption of the 'unrestrained expansionism and resource profligacy that is inherent in the global capitalist project'. An essential antagonism between movements set up to defend the environment and this capitalist project would seem to follow directly. However, the further development of Sklair's position involves him in recognizing both an increasing tendency for a 'greening' of capitalism and widespread co-option and adaptation on the part of green and environmentalist organizations and movements by and to capitalism. These insights might point in the direction of a much more polyvalent view of the 'culture-ideology' and politics of environmentalism in which diverse, opposed social and political interests and value-perspectives engage in struggle over this range of issues. Such a view is foreshadowed in Sklair's concluding discussion of 'shallow environmentalism' as an aspect of the global capitalist system's attempt to resolve its own ecological problems at the expense of Third World countries.

This theme is continued in Chapter 11 by Buttel and Taylor. They begin with some critical comments on key assumptions of social science research, which reflect back upon similar themes addressed by our other contributors: the tendency to accept the nation-state or individual societies as an unquestioned unit of analysis; the failure of theories of social movements to address the very different conditions prevailing in Third World countries; and the ambiguity in much sociology of science, which is sceptical and relativizing with respect to 'establishment' science but uncritically accepting with respect to environmentalist science.

Buttel and Taylor conclude by proposing a positive research agenda for environmental sociology which would overcome these and other difficulties. In doing this they draw upon, and synthesize, a range of sub-disciplines already addressed in other contributions. Sociology of science, the sociology of environmental movements, policy analysis and global sociology. They provide a typology of approaches to the sociology of science, arguing that these are not (or not all) mutually incompatible. Environmental sociology needs to be more influenced by these newer approaches than it has been so far, but, in particular, there is a need to develop a *synthetic* sociology of science which both recognizes science and technology as productive forces *and* considers the cultural and 'ideational' aspects of science. The chapter continues with a line of argument which reflects interestingly on Sklair's concluding pages. The core of their argument is that the 'construction' of environmental issues as global in character has served the interests of both environmental organizations and environmental science. They challenge the prevailing view of the environmental movement as oppositional, and consider the way in which environmental concerns have displaced considerations of social justice in the development-aid establishment. These, and other issues surrounding the way the global environment makes new demands on social theory, are discussed by Shove in her postscript (Chapter 12).

BIBLIOGRAPHY

Adam, B. (1990) *Time and Social Theory*, Cambridge: Polity; Philadelphia: Temple.
Baudrillard, J. (1989) *America*, London: Verso.
Benton, T. (1991) 'Biology and social science', *Sociology* 25(1), 27–49.
Douglas, M. (1985) *Purity and Danger*, London: Routledge.
—— (1990) 'Risk as a forensic resource', *Daedalus* 119(4), 930–55.
Durkheim, E. (1982) *The Rules of Sociological Method* (ed. S Lukes), London and Basingstoke, Macmillan.
Giddens, A. (1979) *Central Problems in Social Theory*, London and Basingstoke, Macmillan.
—— (1981) *A Contemporary Critique of Historial Materialism*, London and Basingstoke, Macmillan.
Harvey, D. (1973) *Social Justice and the City*, London: Arnold.
IIED (International Institute for Environment and Development) (1991) *UNCED: A User's Guide*, London: IIED.
IUCN/UNEP/WWF (1991) *Caring for the Earth: A Strategy for Sustainable Living*, London: Earthscan.
Kemp, D.D. (1990) *Global Environmental Issues: A Climatological Approach*, Routledge, London.
King, A.D. (ed.) (1991) *Culture, Globalisation and the World-System*, London: Macmillan.
McCormick, J. (1989) *The Global Environmental Movement*, London: Belhaven.
Marshall, G. (1982) *In Search of the Spirit of Capitalism*, London: Hutchinson.
Maunder, W.J. (1989) *The Human Impact of Climate Uncertainty*, London: Routledge.
Meadows, D.H., Meadows, D.C. and Randers, J. (1992) *Beyond the Limits*, London: Earthscan.
Newby, H. (1991) 'One world, two cultures': sociology and the environment', BSA Bulletin *Network* 50 (May): 1–8.

Parry, M. (1990) *Climate Change and World Agriculture*, London: Earthscan.
Redclift, M.R. (1992) 'Sustainable development and global environmental change: implications of a changing agenda', *Global Environmental Change* 2(1) March, 32–42.
—— (1993) 'Sustainable development: needs, values, rights', *Environmental Ethics* 2(1): 3–20.
Rose, S., Kamin, L.J. and Lewontin, R.C. (1984) *Not in Our Genes*, Harmondsworth: Penguin.
Schroeder, R. (1992) *Max Weber and the Sociology of Culture*, London: Sage.
World Commission for Environment and Development (1987) *Our Common Future*, Oxford: Oxford University Press.

2

BIOLOGY AND SOCIAL THEORY IN THE ENVIRONMENTAL DEBATE

Ted Benton

In a recent lecture, the chair of the British Economic and Social Research Council, Howard Newby, presented a powerful indictment of the prevailing patterns of research into environmental problems. Natural scientific research, he argued, is overwhelmingly governed by a 'technological determinist' view, according to which natural science and technological changes ('processes') are represented as exogenous to human society. The social sciences are acknowledged to have a place in analysing only the *impacts* of these processes of change, and policy responses to them. Newby argues that there is a clear and urgent necessity for social science, and sociology in particular, to address questions as to the socio-economic and political *conditions* and *causes* of technical changes in relation to their environmental effects. Science and technology are themselves aspects of human society, and we need to understand sociologically how they change and develop.

But, Newby argues, there is a corresponding and symmetrical failing on the side of the sociological research community. Relatively few sociologists have addressed environmental issues, and, where they have done so, they have tended to focus on sociocultural aspects of the rise of environmental movements, and shifts in perceptions of and valuations of the environment. They have not, characteristically, been concerned with developing a sociological understanding of what Newby calls the 'material' aspects of the relation between society and nature:

> The public is broadly aware of the fact that environmental degradation, whether at the global, national or local level, is a result of *human* intervention in natural systems and, in particular, our current patterns of economic development and social organisation which place a burden on the earth's resources which are unsustainable in the long run.
>
> (Newby 1991: 2)

Newby's diagnosis of this state of affairs is that it derives from what C.P. Snow called 'the two cultures'. The deep antagonism and conceptual gulf between natural science and technology on the one hand, and the humanistic traditions in the humanities and social sciences on the other. Several writers – including Redclift (1984), Benton (1991), Dickens (1992) and Newby himself – have pointed out that this weakness of sociology is more than a mere contingent fact about the research interests of its practitioners. Rather, the conceptual structure, or 'disciplinary matrix' by which sociology came to define itself, especially in relation to potentially competing disciplines such as biology and psychology, effectively excluded or forced to the margins of the discipline such questions about the relations between society and its 'natural' or 'material' substate. Redclift and Woodgate, in Chapter 3, illustrate this process in relation to the heritage of Marx, Weber, and Durkheim, the 'founding figures' of modern sociology.

There is now quite widespread agreement that in order to do worthwhile sociological research on the 'material' dimension of environmental issues, the basic conceptual legacy of the sociological traditions has to be radically re-worked. In particular, the dualistic oppositions between subject and object, meaning and cause, mind and matter, human and animal, and, above all, culture (or society) and nature have to be rejected and transcended. The really difficult problems only *start* here, however.

There are three principal reasons why moving beyond this point is so difficult. First, these dualistic modes of thought go very deep. They are not mere superficial devices which can simply be eliminated from the discipline whilst leaving everything else in place. They are in a very important respect *organizing* categories, both shaping sociological thought and research across the whole span of the discipline *and* structuring everyday non-scientific and common-sense contexts of thought. So pervasive are these ways of thinking that the attempt to transcend them is a veritable exercise of pulling oneself up by one's own boot-laces. The second reason is that the sociological traditions' attempts to separate themselves from such other disciplines as psychology and biology were well-grounded. Durkheim was right to insist that human social life is a reality, shaping the lives, relationships and consciousness of individuals. Weber was right to insist that human social interaction is more that mere co-ordinated physical movement. Engels was right to point out that agricultural production marks a way of relating population to environment quite unlike anything to be found in non-human species. If we try to go beyond the culture/nature opposition, it is necessary to avoid simultaneously falling back into the forms of biological or psychological determinism which the founding figures of the discipline were right to oppose.

The third reason why it is so difficult to move beyond the dualism of the nature/culture opposition is that it involves the daunting creative task of developing new concepts for analysing and thinking through the relationships and processes which were previously allocated to their respective conceptual 'boxes' and posted to the appropriate address: 'natural science' or 'social science'. Several of the chapters in this volume are themselves devoted to this creative task, and I, too, attempt to make a contribution to it in a later section of this chapter. However, it might well be worth starting out by identifying 'dualism' in the context of the wider environmental debate, both to see why it is such a pervasive form of thought and to explore some of its limitations.

As we have seen, the dualist strategy of thinking about 'nature' and 'society' (or 'culture') as qualitatively distinct realms offers one obvious and unambiguous way of resisting biological determinism in various fields of sociological analysis. For example, the opposition between sex (biology) and gender (culture) enables us to see how 'masculinity' and 'femininity' differ radically in the way they are 'constructed' in different human cultures, how social relations between men and women might be changed to meet the aspirations of women, and so on. In the well-used phrase 'biology is not destiny'.

In the environmental debate, also, dualistic oppositions between 'nature' and 'society' are an effective way of resisting the analogues of 'biological determinism'. In the discourses of environmentalism we may distinguish two such patterns of thinking – what I will call 'naturalistic reductionism' and 'techological determinism'. Both of these patterns of thinking accord *some* place for social relations and practices in the analysis of environmental issues, but they radically understate, or undertheorize what that place is. Nature/society dualism is a way of resisting these patterns of thought, and insisting that society plays its own independent role, which needs to be analysed and understood. This is very much the position advocated by Howard Newby in the lecture I discussed at the beginning of this chapter. It is an approach which suggests a reasonably clear-cut division of labour between different but complementary scientific disciplines: the natural sciences on one side, the social sciences on the other. It resists naturalistic reductionism and technological determinism by insisting that there is a much more important role to be played by the social sciences in exploring environmental problems than is recognized by those other approaches.

However, as we shall see, nature/society dualism suffers from limitations of its own. Not only this, but it tends to be a rather unstable position, readily 'sliding off' into what might be called 'sociological reductionism'. So long as the natural science ('nature')/social science ('society') division of labour *itself* remains unchallenged, it remains possible (and, indeed, comfortable) for social scientists to bracket off 'nature' as something the natural scientists will deal with while *they* get on with studying the 'social' side of things. Subsequently 'nature' comes to be understood only by way of its cultural representations in the social movements, environmental organizations, or policy debates which are the primary objects of sociological study, and we are back with the type of research which Newby urges us to go beyond. The *interface* between human social practices and their material conditions and consequences is lost to view. This may, as I have suggested, happen as an unselfconscious 'slide' from an explicitly dualist position, but, equally, it may be arrived at as a result of positive reasoning processes. As we shall see, some 'social constructionist' positions are based on the claim that *all* views of nature are symbolic constructs of some culture or other. For this tradition of thought in sociology, Howard Newby's distinction between the environment as a 'set of symbols' and its material aspect is simply not defensible: we can't get outside the symbolic order so as to study the relation to it of external 'nature', considered as it is in itself, independently of human cultures.

For some purposes, it makes sense to think of each of these approaches – naturalistic reductionism, technological determinism, and sociological reductionism – not so much as *alternatives* to nature/society dualism but, rather, as *variant forms* of it. In each case 'nature' is counterposed to 'society', but at the polar extremes one of these opposed terms tends to swallow up the other. In naturalistic reductionism human society is seen as a part of the wider totality of nature, whereas in the more extreme forms of sociological (or 'discourse') reductionism, 'nature' becomes transmuted into its symbolic representations.

In what follows, I'll offer a critical discussion of each of these rival approaches in turn, with the aim not so much of showing that they are *mistaken* but, rather, of using their independent insights as a way of moving beyond what seem to me to be their limitations.

TECHNOLOGICAL DETERMINIST AND 'TECHNOCRATIC' PERSPECTIVES

In their various forms, these patterns of thinking are the primary targets of Howard Newby's criticisms. They are very pervasive in both 'official' and common-sense thinking, and they shape a large part of the environmental research agenda and public debate about the environment. They stem, historically, from the more 'optimistic' versions of the Enlightenment view of 'progress', and have functioned to legitimate a certain view of human development and well-being

31

which has been shared across modern capitalist, state-socialist and 'modernizing' Third World countries. Although, as I suggested above, there are several different variations of this pattern of thought, some themes are identifiably present in all versions. The dominant theme in each case is the idea that there is a long-run, autonomous historical tendency for scientific knowledge to grow and accumulate. Scientific knowledge is itself mainly understood as a means of controlling nature, of making nature serve human purposes. Technological innovation is the medium through which science is applied in this great historical project of growing human mastery of the forces of nature.

In the most influential version of the technological determinist view – sometimes referred to as 'cornucopian' (see Cotgrove 1982), or 'Promethean', or as 'technological optimistic' – there are two aspects of mastery of nature. One is that we should become increasingly able to protect ourselves from formerly catastrophic threats from nature: storms, floods, droughts, diseases, predators, and so on. The other aspect is that nature should become an indefinitely expanding reservoir for the satisfaction of human desire. Science and technology promise an end to poverty, insecurity, and disease, and a prospect of ever-growing material prosperity and cultural enrichment.

Within the 'cornucopian' version itself, it is worth noting that further differentiations can be made. The form most prevalent in industrialized capitalist societies presents a view of the good life in terms of ever-expanding individual choice in consumer goods which satisfy the full range of human needs and desires. Capitalist competition, itself driven by consumer demand, ensures that new scientific knowledge is fully and rapidly utilized in marketable technical innovations. The alternative version of the cornucopian view prevailed in the 'formerly actually existing' state socialist societies. Its view of the good life differed little from the capitalist version, save that it emphasized distributive justice whilst giving less attention to individual choice. However, where the state-socialist version did differ sharply was in its view of the relation between capitalist forms of economic organization and technological innovation. This view sees capitalistic private property and competition as a *constraint* on the further development of human productive powers. The good life of material abundance with the mutual benevolence and conviviality which would flow from it, could, in this view, only be achieved as a result of the destruction of capitalist private property.

State socialism is widely thought to have become discredited among its own citizens, primarily because it failed to deliver in accordance with this promise, by comparison with industrialized capitalism's relative success. Accordingly, among those (especially in the Third World), who continue to share the 'cornucopian' vision of the good life, the Western capitalist variant is now dominant. However, the 'cornucopian' view is by no means universal. From its earliest pre-Enlightenment foundations this vision has encountered deep-rooted opposition. Much of this opposition, whether it thought of itself as conservative in its resistance to modernity, or whether it advocated some alternative vision of the future, tended to emphasize the social and cultural consequences of a secular, 'materialistic', and commercial civilization. Critics were nostalgic for the lost sense of community and shared values, for the courage, nobility of spirit and public virtue which were being abandoned in the scramble for material prosperity or a life of ease and luxury. Crime, corruption, subservience, dull conformity, and self-indulgence would be the 'downside' of a society which would come to justify its central institutions by their capacity to underwrite ever-growing material abundance.

I'll return later to consider the implications of these critical traditions, but for the moment my interest is in another line of criticism of the cornucopian view. This focuses not (directly, at least) on the social and cultural consequences of the growth of science and technological advance, but, rather, on their consequences for nature itself. Versions of the Faust story, the myth of Prometheus, and Mary Shelley's *Frankenstein* can all be read as early warnings of the perils which may come from humanity's arrogance in seeking to master the forces of Nature. Much of our contemporary environmentalist writing shares this theme but now itself adopts the mode of scientific rationalism. Overwhelmingly the most influential text in this contemporary genre has been the Club of Rome's *Limits to Growth* (Meadows *et al.* 1972). A computer model of the world system yielded the conclusion that continued growth in population, agricultural production, industrial production, resource depletion and pollution must lead, sooner or later to global catastrophe.

Interestingly, however, the *Limits* did not itself call into question the cornucopian vision of the good life. Nor, indeed, did it call into question the contribution of science and technology to human welfare. On the contrary, the model developed by the *Limits* team was a specific *use* of scientific method, and their proposals for a steady-state economy, in which growth in production would be contained within limits set by environmentally-improved technologies, opened the way for a 'managerialist' approach to the environment. In this view, further technological innovation is seen as offering solutions to the problems posed by earlier waves of technical development and industrial growth. New, and cleaner, technologies, and an emphasis on quality of life and provision of services, as against the older forms of 'smoke-stack' industry and mass production of material goods, come to form the new image of the 'good life'. Notwithstanding this shift, the newer forms of environmental managerialism retain the basic commitment to technological determinism which characterized the increasingly discredited cornucopian view. Science and technology continue to be viewed as having, so to speak, lives of their own, autonomous dynamics of discovery and application, the pace of which, for ever accelerating, sets the agenda for social and economic change, the adverse consequences of which must be scientifically monitored and expertly managed.

The core assumptions shared by all these variants of the technological determinist view are that there is a single-line cumulative growth of scientific knowledge in history (or, at least, since the scientific revolution of the seventeenth century), and that this knowledge gives rise to progressive mastery of nature through its application in technology. A further secular-materialist value standpoint is also generally present: this is that human well-being, the 'good life' consists in the ever-growing gratification of human desires by way of this technologically mediated mastery of nature. Well-grounded work in the history and sociology of science and technology raises serious doubts about these core assumptions, whilst there are good sociological and philosophical arguments for doubting both the claim that modern science and technology do generally achieve the levels of want-satisfaction they promise, and the adequacy of the secular-materialist view of the good life (at least in its technological determinist forms).

In the English-speaking world, the work of Thomas Kuhn (especially *The Structure of Scientific Revolutions*, 1962, 1970) played a crucial role in calling into question the then prevalent view of the history of science as a straight-line process of continuously accumulating objective knowledge about nature. He offered persuasive analyses of revolutionary episodes in science which seemed to show that they involved radical qualitative shifts of perspective, in which previous knowledge had to be *discarded* as well as built upon. Mere logic and empirical evidence were insufficient to determine the outcome of such episodes, so that any full understanding of scientific change would have to take into account social processes – including power relations within the scientific community. Moreover, Kuhn's investigations yielded evidence about the extent to which external pressures on science – the requirement for calendar reform, or for more adequate understanding of the mechanics of projectiles – played a part in precipitating crises of legitimacy for established scientific theories, and in setting criteria of acceptability for their rivals. Kuhn's historical understanding of science as a social practice opened up the possibility of a sociology of science in which the *content* and *direction* of scientific change could be investigated. Many sociologists of science would go still further, and argue that with the development of modern industrial capitalism (and state-socialism) the research agenda of the natural sciences has become more and more integrated with the technical and marketing requirements of industry and the military.

Political, economic, and military interests shape, by way of the organization and funding of research, the research priorities and the formulation of problems for investigation on the part of the scientific 'community'. Even 'basic' science, it could be argued, relates to these interests, albeit in a more open-ended way and with a longer time-scale in view. Moreover, in the last twenty-five years or so, historians and philosophers of science have shown the extent to which theoretical innovation in science is a creative process, going well beyond the fitting of evidence into generalizations. The role of analogy and metaphor in constructing models of natural mechanisms and processes is now widely acknowledged to be an indispensable feature of scientific thinking. Darwin's concept of 'natural selection', for example, embodied a metaphor drawn from the selective breeding of domestic species. Another example from biology is the concept of sequences of DNA as a 'genetic code'. As these examples show, scientific thinking is dependent upon the availability, within the wider culture of the scientists, of social and cultural practices which can serve as sources for metaphorical thinking.

These sociological considerations point away from a view of science as autonomous and linear-cumulative in its historical development. Instead, they suggest that scientific knowledge will be to some extent, at least, shaped by its specific, and possibly localized cultural context, and by the constellation of social, political and economic interests which fund research and govern its institutional form. Similar considerations can be brought to bear on the development of technology. Whilst it is true that a great deal of scientific research is funded with a view to its possible application in marketable or usable technologies, science cannot be justifiably *reduced* to this motivation. Obviously, there would be no point in expenditure of money and effort on scientific research if its outcome could be predicted in advance. Many technically usable discoveries have been unexpected 'by-products' of research conducted for other purposes. Many discoveries remain for long periods without anyone recognizing that they can be applied in technology. Also, a great deal of scientific knowledge concerns processes and mechanisms in nature which are beyond the reach of (foreseeable) human technical manipulation. Much astronomical research has this character, as also does a great deal of research in evolution and palaeontology. Darwin's discovery of the mechanism of organic evolution had a profound cultural, moral and metaphysical significance, but added nothing to our technical mastery of nature.

Widely shared views of science (including those of Popper, logical empiricism, 'instrumentalism', and many of the thinkers in the Frankfurt tradition of critical theory such as Adorno, Horkheimer, Habermas) which make close logical ties between science and its application in technology fail to acknowledge the extent to which patterns of social and economic interest and power govern *which* scientific discoveries get applied in technology, and, also, *which* technologies get constructed. For example, sociological research on labour processes and technological innovation in industry suggests that the emphasis is on innovations which de-skill and replace human labour, and enhance management control and predictability of production. It is not hard to imagine that in a society marked by consensual, rather than antagonistic, industrial relations, in which producers themselves set the priorities for technical innovation, technologies might develop in quite different ways. The emphasis might be, for example, the fostering of more stimulating and varied work tasks, on convivial group-working, and on more congenial working environments.

Moreover, for any given technology – the microchip, the steam engine, genetic manipulation – there will be in principle a whole range of 'options' about the specific range of products or commodities which they are used to make, about the place they come to occupy in social life, and whose and which needs they come to satisfy. I put 'options' in inverted commas, since, in effect, in contemporary societies concentrations of economic and political power close off any sense of a public debate or communal decision-making about these issues. It is, indeed, this foreclosure of debate, and the absence of communal control over scientific research funding and technical innovation, which gives the *appearance* that science and technology are autonomous, all-determining forces shaping society. The technological determinist picture is false, but it is made *plausible* by an inscrutable concentration of power over science and technology in both advanced capitalist and state socialist societies.

The 'cornucopian', or 'technological optimist', version of technological determinism postulates a potentially limitless capacity for technological innovation to secure mastery over the forces of nature, and to resolve such obstacles as arise in the way of this project – most notably the ecological damage done by the Promethean project itself.

This can be – and, indeed, *has* been – presented as a deeply risky wager[1] on the future of ourselves and our planet – one which the technological optimists have so far won. However, *at best* there are no grounds for supposing that we are not in the position of the proverbial man, falling from a 20-storey building, who was heard to call out 'so far so good' as he passed the sixth floor. There are numerous potentially deeply threatening environmental problems – from the safe disposal of nuclear waste through to global climate change – for which 'technological fix' solutions are not remotely in sight. Still more seriously, however, posing the solution to environmental problems in terms of the invention of technologies directs attention away from gross global inequalities in power and resources which allow tens of millions, especially in Third World countries, to suffer and die as a result of ecological destruction. Desertification, lowland flooding, over-cultivation of marginal land, water pollution and drought, industrial disasters such as Seveso and Bhopal, are processes and events which kill and maim on a horrifying scale already. The 'limits to growth' have already been reached for many human populations, but this is not because of a lack of scientific or technical know-how.

Our other variant of technological determinism is hardly less questionable. The 'managerialist' approach, pioneered by the Club of Rome and still very much present in the 'Northern' environmental agenda at the recent 'Earth Summit' at Rio de Janeiro, wears an eco-friendly mask: it, at least, explicitly conceptualizes objective, nature-given limits to human expansionary appetites. Where these latter cannot be sustainably met, they must be controlled and regulated.

This form of technological environmental ideology is open to criticism at several levels. First, because it shares with other variants the basic thesis of technological determinism it can only define and propose solutions to environmental problems as they are produced and encountered by the already dominant paradigm of growth and development. In other words, it can postulate reduced growth, or even stasis, but it cannot contemplate qualitatively different lines of sociocultural and economic change. This is of great political significance. If the science and technology we have, and are likely to get, are products of a particular set of cultural priorities or economic-political interests, then the 'limits to growth' are constraints on the global projection of just those particular interests and priorities. In this light 'environmental management' can now be seen as a global strategy for securing and regulating the conditions for the long-term sustainability of a *particular kind* of human culture and its dominant economic and political interests. So, the protests that were voiced, however hypocritically, by some Third World political elites, that the Northern environmental agenda at Rio amounted to a new form of imperialism may not have been so wide of the mark.

Second, the 'environmental management' perspective has tended to radically undertheorize the social, legal and political processes of environmental regulation themselves. Models of sustainability tend to take the form of equilibria between human-generated demands on the environment (in terms of population, pollution, resources, etc.) and the 'carrying-capacity' of the global environment with respect to these demands. However, almost entirely absent from such models are sociologically informed discussions about what kind of institutional framework would be required to maintain such equilibria. In the absence of serious theorizing the reader is left to suppose that existing economic structures, power-structures, and legal/political institutions would remain broadly in place but would be given a new set of policy priorities: ministries would develop energy, transport and industrial (etc.) policies within agreed environmental constraints, businesses would be given economic incentives for ecological good conduct or abide by environmental legal restraints, whilst foreign policy would be oriented to reaching responsible international agreements on transnational environmental issues. Of course, these hidden assumptions have only to be spelled out for their sociological implausibility to become evident.

In the context of a world in which there are already massive global inequalities in material well-being, and also growing inequalities within individual nation-states, it is to be expected that local, regional and global strategies for environmental management will be confronted with conflict between those who, on the one hand, stand to lose out in terms of profits, jobs and livelihoods, and, on the other hand, stand to lose most (and most immediately) as a result of environmental destruction. The perspective of environmental management, tied, as we have seen, to technological determinist assumptions, is not amenable to a democratic or consensual resolution of such interest conflicts in society. An increasingly authoritarian imposition of a science-based strategy would seem to be the most likely scenario for this way of thinking. Indeed, the systems approach which has been its major intellectual influence is deeply sceptical of the possibility that democratic processes could perceive and adequately respond to system problems.

NATURALISTIC REDUCTIONISM

Technological determinism, and technocratic ways of thinking, are, of course, not the only ideologies which understate the role of society in the environmental debate. Often to be found at the opposite pole of the political spectrum in radical, or 'deep' green circles, are patterns of thinking which wholly reject the value-orientations typical of the technological determinist view of progress and development. But these are rejected in favour of a mode of human life for which nature itself provides the model. Though there are, again, many variant forms, there are characteristic metaphysical beliefs and threads of argument. Humans tend to be conceptualized as a species of natural being, living alongside, and interacting with others. As with other species of living organism, we have a mode of life which involves a range of characteristic interactions with, and dependencies upon, our bio-physical environment. We are *part* of nature, not set over and against it. We are connected to other elements in the biosphere by a seamless web of interdependencies. On this view the concepts of ecology as a biological science apply in an unqualified way to the human species, whilst the philosophical principles underlying ecology are generalizable as a set of norms for human conduct.

Malthusian writers abstract from socially produced differentials in the ecological impact of different human populations, and reduce human ecology to the equation of population to resources. This is one form of naturalistic reductionism, which has close affinities to sociobiology's attempts to reduce human social interaction to the expression of 'selfish genes'. However, it is a form of reductionism not often to be found in the radical wing of the 'green' movement. Much more widespread are versions which do acknowledge general-ized features of human social organization as having a causal role in environmental degradation. These deep-green perspectives often rely on some version of an arcadian 'golden age' in which humans lived in harmony with one another and with nature. Subsequently things have gone profoundly wrong, both within society and in human rela-tions to nature. Just *what* has gone wrong is a matter of controversy. In some versions (for example, the social ecology of Murray Bookchin and Janet Biehl) it was the establishment of social hierarchies, and the subsequent development of capitalism and the nation-state as forms of domination of both humanity and nature. In other versions the imposition of patriarchy on earlier harmonious and gender-equal societies was the beginning of a historical process in which a masculine project of simultaneous domination of women and nature is respons-ible for the runaway ecological destruction now confronting us. Still other versions put the burden of blame on pervasive forms of con-sciousness and value orientations: those derived from 'mechanical science', 'atomism', 'dualism', 'anthropocentrism', 'logocentrism', without asking too many questions about the forms of human society in which these ways of thinking flourish.

Despite the variability of their diagnoses, these perspectives are remarkably consistent in their versions of the 'cure': a return to a materially more simple, egalitarian and convivial, decentralized communal existence. Such a society would replace the endless and destructive scramble for worthless commodities by the pleasures of social communication and participation, and an enhanced spiritual and aesthetic connectedness with nature. For some, the achievement of such a society would constitute a return to the values of the long-lost 'golden age', whilst for others it would constitute a further stage in human progress – a dialectical resolution of the ecological and social contradictions of our past.

It is a fairly straightforward matter to show where the Malthusian versions of naturalistic reductionism go wrong. The massive differences in per capita pollution impact or resource-use as between industrialized and Third-World countries (a ratio of 40:1 is an often-quoted estimate) clearly put the primary *causal* responsibility for global environmental degradation at the door of the industrialized countries rather than on the poverty-stricken populations of the South. This becomes even clearer if we shift away from the sociologically rather suspect measure of 'per capita' impacts and consider, instead, the relations of power characterizing the world economic system and the sources of the growth dynamic in the First-World-based transnational corporations and financial agencies. However, even the dynamics of population growth cannot be adequately understood in Malthusian terms. The rates at which populations *actually* grow is not a function merely of numbers, ages and fecundity, but relates closely to such social-relational features as property ownership, women's rights, social security provision for the aged, health services and educational provision.

These weaknesses in the Malthusian approaches can be seen as a consequence of their naturalistic reductionism. In this case, the reductionism consists in supposing that the characteristic features of human interventions with their environments can be analysed in terms taken, unmodified, from scientific ecology. We can, and, I think, we should, continue to view humans as a species of living organism, comparable in many important respects with other social species, as bound together with those other species and their bio-physical conditions of existence in immensely complex webs of interdependence, and as united, also, by a common evolutionary ancestry. To say this much *is* to be committed to a naturalistic approach, but not necessarily to a reductionist one. It is to be committed to recognizing the *relevance* of evolutionary theory, physiology, genetics and, especially, ecology itself, as disciplines whose insights and findings are pertinent to our understanding of ourselves.

However, there are strong reasons for thinking that humans constitute a 'special case' in two respects – first, that these life sciences are *insufficient* of themselves for an understanding of human personal and social life. This is because humans have evolved emergent powers which demand new and distinctive modes of analysis. Second, humans are a special case in that acquisition of these emergent powers also qualifies, or modifies the *ways* in which the other life sciences are relevant to us. So, the concepts of ecology, for example, whilst *applicable* to us, are not applicable in an unqualified or unspecified way.

So, what are these 'emergent powers'? Notwithstanding their close kinship with other primate species, humans do seem to be distinctive both in their biological organization – most notably their brain, central nervous system, and vocal organs – and in their *forms* of sociability. In evolutionary terms these two features appear to be connected. The quite unique flexibility in human social co-ordination of activity is made possible by symbolic communication. This is conventional in character, acquired by individuals through social learning, and specific in its content to particular socially interacting groups. This feature is most commonly recognized as the distinctive human capacity for language. Whilst I accept that language is, indeed, paradigmatic of symbolic, conventional communication, I would prefer to distance myself from this way of characterizing human distinctiveness. For one thing, it is quite clear that some of our closest primate kin *can* acquire many of the capacities which go to make up linguistic ability in the human case. For another, I would want to emphasize the extent to which linguistic communication takes place only in the context of forms of human life and practice in which other, non-linguistic symbolic devices (facial expressions, gestures, postures, embraces, and the like) are also available and may take primacy. More pertinently, however, I want to put the emphasis on the place of symbolic communication in the co-ordination of social practice as a key feature of our natural history as a species.

A further feature, very closely connected with the above, is our individual capacity for moral agency: that is to say, to regulate (or refuse to regulate) our activity in accordance with normative rules or principles. This is the correlate at the level of the human individual, of the characteristic feature of human societies that they integrate the activities of their numbers through processes of normative ordering. These distinctive features of human social life – symbolic co-ordination of activity, together with moral agency and normative regulation – give human social practices their unique flexibility. Collective learning and reflexive monitoring render them susceptible to intentional modification and conscious adaptation in a way which is quite unique to our species.

But if we continue to bear in mind that this species still remains an organic species – embodied, sexually reproducing, subject to the organic requirements of food, shelter and (generally) clothing – it is clear that our uniqueness in these respects by no means lifts us out of the order of nature. We remain both unavoidably organically embodied and ecologically 'embedded'. However, human distinctiveness does have a number of corollaries of direct relevance for our

TED BENTON

understanding of human ecology. There are three which I particularly want to emphasize here:

1 The combination of symbolic communication and normative regulation of activity is both a distinctively human set of *capacities* and, at the same time, a distinctively human need. Lacking what Mary Midgley has called 'closed instincts', humans depend for their capacity to identify and meet their full range of needs upon the conceptual resources and normative rules which constitute their local culture. Since that culture is itself at least in part constituted by conceptualizations of the non-human environment and normative rules governing interaction with it, human ecology is characterized by a complex and interwoven set of interdependencies between ecological conditions and contexts of life, social and cultural forms, and personal identity and well-being. Humans are, therefore, *vulnerable* to environmental degradation and dislocation in a multitude of ways, some of them quite peculiar to the species. Specifically cultural, identity, self-realization and aesthetic needs interact with and complement organic needs for food and shelter in ways which figure less, if at all, in the ecological requirements of other species.

2 But, just as there are so many more ways in which human relations with their environments can go wrong, it is also the case that humans have a quite unique ability to enhance the carrying capacity of their environments for populations of their own species (or, incidentally, for other species if they so choose). To some extent it is true that other species maintain their habitats by their own activities. Grassland herbivores, for example, by their own grazing activity, prevent the scrubbing-over and hence reduction of their grassland habitats. However, humans are arguably unique in their combination of capacities for extending their sensory and motor powers *vis-à-vis* their environments by way of invention of tools and weapons, cognitive 'mapping', domestication of other species, and large-scale social co-ordination of activity. Though there are some striking analogues of particular items on this list – the social co-ordination achieved by the social *Hymenoptera*, the dam-building of beavers, the nest-building of birds, the termite-fishing of chimps and the navigational abilities of migratory species – no other species shares anything like this overall *pattern* of capacities as an intrinsic feature of its natural history. Arguably, again, this distinctive pattern is a corollary of symbolic communication and normature ordering as human-distinctive emergent powers. Human inventiveness, with respect to our powers of intentional modification of our environments through normatively ordered social

42

practices, renders quite illegitimate any attempt to read off from a specification of the bio-physical environment what its 'carrying capacity' might be for human populations. The concepts of ecology in their application to the human case must be crucially qualified to take account of the cultural and historical variability of human social practices of environmental regulation and transformation.

3   It follows that there is no 'natural' mode of human relation to nature. No original, ecologically 'harmonious' golden age or state of grace from which we have fallen. Humans have no single, instinctually prescribed mode of life, but a range of indefinitely variable 'material cultures'. The ecological consequences and conditions of human/environmental interaction are a function of each *specific* mode of social life in relation to its ecological sustaining conditions and bio-physical media of activity. Each form of society available for anthropological study is characterized by its own *specific* constellation of limits, affordances and vulnerabilities to ecological unintended consequences. The forms of human ecology, as culturally mediated relations to physical, chemical and biological conditions, are both limitlessly variable *and* ecologically bounded.

The bearing of this line of argument upon some radical green forms of naturalistic reductionism should now be fairly obvious. Although a good deal of cultural anthropology has been characterized by just that problematic opposition between nature and culture which I have been criticizing in this chapter, there is an increasingly significant body of anthropological evidence to indicate that many pre-industrial cultures have encountered severe and self-generated ecological problems, and, in some cases, these seem to have played a significant part in the fall of major ancient civilizations. There is no 'golden age' to which we can look back for a model of our own ecologically sustainable future.

Further, it is a mistake, which many Greens borrow insufficiently critically from the technological determinist tradition, to think of 'Nature' as a finite system of constraints which sets outer limits to human demands. Andrew Dobson, for example, takes this as definitive of 'green', as distinct from 'environmentalist' politics:

> it is often forgotten that the foundation-stone of Green politics is the belief that our finite Earth places limits on our industrial growth. This finitude, and the scarcity it implies, is an article of faith for Green ideologues, and it provides the fundamental framework within which any putative picture of a green society must be drawn.
>
> (Dobson 1990: 73)

The line of argument I have just developed suggests, rather, that any concept of ecological 'limits' or 'boundaries' must be relativized to particular social and cultural forms, which, in turn, should be thought of as employing some definite range of natural mechanisms as their sustaining conditions and media. The cultural mediation of human social activity in relation to the non-human environment is such that there may be an indefinitely large number of *qualitatively different* such forms, each with its own dynamics, each with its own particular pattern of ecological obstacles and vulnerabilities.

It is entirely feasible that there may be numerous possible, but qualitatively distinct, directions for future sustainable development. Each will have to observe its ecological boundary conditions, but there is no necessity that any will require a return to rustic simplicity, material deprivation or narrow-minded localism – still less to the worship of earth-goddesses! Not only may there be no *single* line of sustainable development, but also it is increasingly clear that sustainable development is itself a concept whose definition presupposes social and cultural preferences and priorities. In other words, green politics will have to define and argue morally and politically for its vision of the future. It cannot coherently rely on the idea of fixed, scientifically definable outer limits as an objective foundation for its vision. Nor can an account – whether drawn from ecology or ecological philosophy – of the properties of nature serve as an objective foundation for a sociocultural ethic. 'Interdependence' can be used to justify hierarchy and stasis, no less than egalitarianism; nature may be viewed as a symbiotic system, or as red in tooth and claw: rival human cultural and political traditions are as much at work in *constructing* these views of nature as they are in drawing congenial lessons from them.

OVERSOCIALIZED VIEWS OF
HUMANITY AND NATURE

Much of the above argumentation seeks to show that the most common patterns of thinking in the environmental debate are radically undertheorized with regard to the economic, cultural, social and political aspects of these issues. However, the clear risk attaching to this line of argument is that it may succeed in demolishing the case, carefully built up by both environmentalists and Greens, that our civilization is profoundly threatened by the ecological consequences of its own patterns of growth. I have tried to avoid this danger by explicitly building my account of human sociocultural (and therefore ecological) distinctiveness into a wider framework which acknowledges both organic embodiment and ecological 'embeddedness' as central and unavoidable features of human natural history.

However, it remains to be discovered whether this approach is effective against those pervasive forms of nature/society dualism which proceed from a valorization of sociocultural life, and yield a perspective in which the independent presence of the non-human world in our lives is marginal to the point of disappearance. The intellectual strategies which show this tendency are very heterogeneous and also very widespread in the human social sciences.

The 'classical' sociologies deriving from Weber and the German neo-Kantian tradition, and from Durkheim and French structuralism, establish (albeit by different theoretical moves) sharp nature/society dichotomies. So, too, do the American tradition of social anthropology and the American sociological traditions deriving from pragmatist philosophy and symbolic interactionism (paradoxically, some of the key 'source' texts for both these traditions were explicitly pitched *against* dualism). Strictly speaking, accepting a nature/culture dichotomy does not entail an oversocialized view of nature, or of nature/society relations. However, the options available are limited: *either* sociology is confined to a domain in which symbolic forms and individual or 'collective' representations can be studied more-or-less in abstraction from their material supports or effects, *or* it is widened in scope to include human embodiment, disposal over material resources, and so on. If the latter, then nature/culture dualism constrains theorizing in ways which tend towards a reduction of the material world to its symbolic investment within human cultures. A subtle shift takes place from the analysis of relations between social life and its material conditions and media, to a cultural analysis of the conceptual frameworks and valuations through which the society under consideration thinks and 'lives' its relation to those conditions. This is, of course, an absolutely indispensable moment, or aspect, of social analysis, but it remains insufficient in so far as it is unable to grasp the ecological and social consequences of *unacknowledged* conditions of social practices in relation to nature, and their unintended or unforeseen consequences.

This 'idealist' tendency in modern social theory has been intensified by the recent 'linguistic turn'. Extreme relativism in the philosophy and sociology of science (Feyerabend, the 'strong programme' in the sociology of science) has interbred with a pervasive misreading of the Saussurean thesis of the arbitrariness of the relation between sign and signified and Nietzschean moral and epistemological nihilism to produce luxuriant growths of hyper-idealism. At most the notion of a reality external to discourse is acknowledged as an unknowable ghostly presence. Keith Tester provides us with a wonderfully entertaining sample of this way of thinking:

Mary Midgley believed that a fish is always a fish, but she was wrong. A fish is only a fish if it is socially classified as one, and that classification is only concerned with fish to the extent that scaly things living in the sea help society define itself. After all, the very word 'fish' is a product of the imposition of socially produced categories on nature. Writers like Lorenz, Wilson and Midgley are wrong: animals are indeed a blank paper which can be inscribed with any message, and symbolic meaning, that the social wishes.

(Tester 1991: 46)

Here, however, I want to concentrate on just two examples of 'over-socialized' approaches. One of these is the 'social problems' approach to environmental issues. In his book, *The Green Case*, Steven Yearley quotes the approach pioneered by Kitsuse and Spector, who argue not only that the existence of 'objective social conditions' is insufficient to explain the emergence of a social problem but also that they are not even necessary:

Kitsuse and Spector have taken the lead in arguing that sociologists concerned with social problems should suspend any interest in whether the objective circumstances merit the existence of a social problem or not. . . . Instead, they should focus on the social processes involved in bringing an issue to public attention as a social problem.

(Yearley 1991: 50)

It is, of course, both interesting and important to be aware of these social processes, but in the case of environmental issues this approach has the consequence of bracketing out of sociological analysis any consideration of the 'objective conditions' which give rise to environmental concern. It is all the same, as far as the sociologist is concerned, whether we do, in fact, face ecologial catastrophe, or whether environmentalists have conjured this threat out of their fevered imaginations. Under the guise of a methodological division of labour between the concerns of sociologists and natural scientists, the type of approach advocated by Kitsuse and Spector effectively excludes the environmental issues themselves from investigation.

Another highly relevant case of 'oversocialized' thinking is provided by Fred Hirsch's very influential argument in his book *Social Limits to Growth*. In several respects, Hirsch's argument lends support to the position I have been trying to develop in this chapter, but Hirsch's conceptual framework makes this difficult to see. One of Hirsch's aims is to explain why it is that continued economic growth leads to both disappointed expectations and increased, rather than decreased,

46

demand for redistribution. He shares the scepticism of other economists concerning the 'limits to growth' thesis. In his view, the case that there are absolute physical outer limits to growth, deriving from finite stocks of raw materials, or of cultivable land, is 'not proven':

> the concern with the limits to growth that has been voiced by and through the Club of Rome is strikingly misplaced. It focuses on distant and uncertain physical limits and overlooks the immediate if less apocalyptic presence of social limits to growth.
>
> (Hirsch 1977: 4)

The focus of Hirsch's study is on the significance of these social limits. In his view, it is a feature of economic development that, as basic material needs are met, the purely private aspect of even individual consumption is displaced by a social aspect. That is to say, the satisfaction to be derived from individual consumption is increasingly affected by the 'surrounding conditions of use'. It is this feature of consumption which imposes social, as distinct from physical, limits to growth, through the operation of what Hirsch calls 'social scarcity':

> social scarcity is a central concept in this analysis. It expresses the idea that the good things of life are restricted not only by physical limitations of producing more of them but also by absorptive limits on their use. Where the social environment has a restricted capacity for extending use without quality deterioration, it imposes social limits to consumption.
>
> (Hirsch, 1977: 3)

Hirsch develops a categorization of social scarcity which acknowledges that it derives from a variety of sources. So, for example, he notes that values attached to some goods, such as collectable antiques, may derive entirely from their scarcity and not at all from their intrinsic characteristics; he calls this 'pure' social scarcity. In other kinds of case, satisfaction is derived from the intrinsic characteristics of the good but is limited by the surrounding conditions of use. He calls this 'incidental' social scarcity. An example is educational achievement, considered as a means to better job opportunities or to social status. This is a good which can be enjoyed only if others do not achieve an equivalent educational standard (assuming constancy of job opportunities and an unchanging status hierarchy). This is a kind of good which offers progressively less satisfaction as more people acquire it. The problem here is what Hirsch calls 'social congestion'. But he also recognizes as a kind of social scarcity the loss of satisfaction to be derived from private car use as more people own and use cars – though in this case consumption limits derive from physical congestion.

Goods which, in these various ways, offer less satisfaction as access to them becomes generalized are termed 'positional goods'.

Clearly, Hirsch's arguments do have a profound bearing on the questions which Greens and environmentalists themselves pose in relation to aggregate economic growth as a measure of human welfare and satisfaction. But the significance of his argument is in part disguised by his continuing to think within a dualist opposition between natural (or 'physical') limits and social ones. Certainly there are some sources of social scarcity – most obviously positions of leadership, power, or high social prestige – which can reasonably be regarded as 'purely' social in character. However, many kinds of positional goods cannot be understood in this way. If we consider, for example, Hirsch's category of social scarcity deriving from 'physical congestion', it is clear that it is capable of further analysis.

As such, car ownership has become less enjoyable as a status provider as more and more people have come to own cars. However, status may still attach to ownership of *specific models* of car which are scarce or expensive. Though, arguably, 'status' as an attribute of social scarcity is 'purely' social, it remains unclear that it is a limit to growth. There seems to be no reason in principle why marketing strategies and the cultural systems in which they are embedded might not go on indefinitely transferring status value from one commodity to another. The mechanism is a familiar one, and it is a mechanism of market growth rather than a limit.

However, increased car use does reduce the satisfaction to be gained from ownership in other respects. Traffic congestion means that journey times are prolonged and rendered unpredictable, parking becomes difficult at the end of journeys, and so on. This does look like a genuine 'limit', but why does Hirsch insist on calling it a 'social' limit? Clearly, it is 'social' in the sense that it is the spread of the social practice of car use which causes the congestion, and the frustration experienced is also at least in part due to the disruption of social expectations and purposes caused by delay and unpredictability. However, it is equally clear that these problems are only experienced because of the characteristics of the spatial distribution of the social practices concerned and because of the physical 'inelasticity' of their physical conditions and means: motorways are of finite width; they cannot serve their purposes without road-junctions; vehicles are subject to break down: and so on. Moreover, patterns of segregation of working life, residence, and shopping or entertainment facilities are themselves consequences of cultural and social processes (including the assumption of generalized car ownership) which nevertheless impose *physical* constraints on individuals seeking to meet their full range of social needs. In short, the situation is one of a tightly

48

intermeshed pattern of socially, culturally and physically bounded opportunities and constraints.

Of course, it could be argued that the physical constraints are not as inelastic as I have suggested. More lanes can be added to motorways; by-passes, under-passes and so on can be built; and cars can be designed more and more effectively to reduce mechanical failure or driver error. This is a classic 'technical fix' response to congestion, and expresses the prevailing 'technological optimist' perspective. However, several points need to be noted. First, these innovations are not cost-free: labour time, physical materials and land are diverted from other purposes, not in order to improve human welfare but to offset a self-induced deterioration. Second, so long as the growth dynamic in car use continues such 'solutions' are highly provisional. In so far as their effects are ameliorative, they increase demand for yet more road space. Third, these by-products of generalized car use have consequences not just for the enjoyment of the benefits of car use itself but for *other* individual and social purposes. CO_2 emissions contribute to global warming, exhaust fumes are unpleasant and unhealthy, urban space becomes irrational and impenetrable for pedestrian use, whilst in the countryside landscape values, ecological diversity and tranquillity (often primary aims of the car users themselves) are destroyed, and so on.

Sometimes, in the environmental debate, we refer to these as forms of environmental degradation. From an ecocentric point of view this may be all that needs to be said. However, it is also the case that these environmental effects of one generalized human social practice in turn adversely affect our capacity for carrying on or enjoying *other* social practices, for satisfying *other* desires and needs. The interweaving of these social, cultural and ecological conditions and contexts of action is such that any simple distinction between 'social' and 'natural' (or 'physical') limits becomes unsustainable in this kind of case. 'Nature' does not function as an absolute outer limit to growth, but nor, in general, do human social relations constitute a limit independently of their interweaving with bio-physical processes and causal mechanisms. We may expect ecological constraints to operate not by way of some sudden catastrophe but through a steady, multi-dimensional dislocation of social practices and frustration of human purposes.

Above all, it is the technocratic, environmental-managerialist perspective, paradoxically enough, which immobilizes policy-making in the face of these dangers. This is for two related reasons. First, the technological determinism implicit in it marginalizes to the point of exclusion the possibility of qualitatively different forms of social, political and economic relations, whose intertwining with ecological conditions might yield quite new possibilities for human flourishing.

Second, technological determinism at best reflects, and at worst actively sustains, the sequestering of scientific and technical innovation and patterns of industrial growth from the arena of public debate and democratic communal decision-making. The industrialized capitalist regimes, and the remaining state-socialist ones despite their global reach, are clear anthropological peculiarities in the absence of such communal or normative regulation of their relationships to nature. The plausibility of the technocratic approach is a result of the coincidence of its assumptions with a widely shared public experience of the forces of technical and scientific advance as demonic forces, out of control and unstoppable. One of the great benefits of a sociological perspective is that it enables us to recognize that what is 'out of control' is not some mysterious *telos* of history, but the key institutional processes of corporate control, state power, and scientific innovation.

NOTE

1 I'm referring to the widely publicized wager between P.R. Ehrlich and J.L. Simon. This was entered into in 1980 and involved predictions of the prices of five metals by 1990. Contrary to Ehrlich's expectations they fell, and he had to pay up $576.07! (*Guardian*, 21 December 1990, p. 25).

BIBLIOGRAPHY

Benton, Ted (1991) 'Biology and social science: why the return of the repressed should be given a (cautious) welcome', *Sociology* 25(1), 1–29.
Cotgrove, S. (1982) *Catastrophe or Cornucopia: The Environment, Politics and the Future*, Chichester: John Wiley.
Dickens, P. (1992) *Society and Nature: Towards a Green Social Theory*, Hemel Hempstead: Harvester Wheatsheaf.
Dobson, A. (1990) *Green Political Thought*, London: Unwin Hyman.
Hirsch, F. (1977) *Social Limits to Growth*, London and Henley: Routledge.
Kuhn, T.S. (1962 and 1970) *The Structure of Scientific Revolutions*, Chicago: University of Chicago Press.
Meadows, D.H., Meadows, D.L., Randers, J. and Behrens, W.W., III (1972) *The Limits to Growth*, New York: Universe Books.
Newby, H. (1991) 'One world, two cultures: sociology and the environment', BSA Bulletin *Network*, vol. 50 (May): 1–8.
Redclift, M. (1984) *Development and the Environmental Crisis*, London and New York: Methuen.
Tester, K. (1991) *Animals and Society: The Humanity of Animal Rights*, London and New York: Routledge.
Yearley, S. (1991) *The Green Case*, London: HarperCollins.

3

SOCIOLOGY AND THE ENVIRONMENT

Discordant discourse?

Michael Redclift and Graham Woodgate

Strategies to achieve sustainable development are situated in structural contexts, economies and societies, in which individual and group interests often diverge. Their success depends on the efficacy of the intentional acts of human agents. In seeking answers to these issues of human agency and social structure we are necessarily drawn towards the discipline of sociology. However, sociology evolved to explain the social realm, and paid scant attention to wider environmental considerations. Traditional approaches to sociological investigation, drawing on the ideas of Durkheim, Weber and Marx, paid relatively little attention to the environmental limits within which human behaviour can be understood. Living within these limits is the major challenge facing us in the twenty-first century.

In this chapter we seek to demonstrate that analytical approaches to the environment can be enriched by paying attention to the tradition in which classical sociology was forged, especially the tension between structural explanations of change and those based on theories of human agency. In particular we examine the theoretical debate within contemporary sociology and seek to demonstrate its utility for the investigation of environmental issues. However, we also believe that sociology needs to embrace not only changes within human society but also the ability of societies to manage and exploit nature. With this in mind, we discuss the concept of coevolution, and suggest ways in which it can help us understand the relationship between the environment and society in both the countries of the North and those of the South.

We can begin with the intellectual inheritance represented by the 'founding fathers' of modern sociology, amongst whom Emile Durkheim remains one of the most challenging and perhaps the least understood. Durkheim distinguished the social realm from the psychic, biological and mineral realms, but he also considered society to be a phenomenon of nature (Durkheim, 1982). This suggests that, in Durkheim's view, nature is both a precondition for society and

51

separate from it. Unlike some modern sociologists Durkheim did not insist that nature was 'socially produced'. Had he done so it would have been considered a 'social fact', imposing external constraints on the individual, and therefore amenable to 'objective' sociological investigation. The Durkheimian emphasis on explaining social behaviour in essentially social terms has served to inform modern sociology in several ways, but this tradition does not easily accommodate the view that our environmental problems are the outcome of the way we view nature. However, Durkheimian approaches do lend authority to the view that environmental problems can only be understood within the context of cultural production and reproduction. The legacy of Durkheim leads us away from simplistic theories of nature/society relations towards a more sociologically grounded view of human 'nature' itself.

The legacy of Max Weber's sociology is rather different. Weber's deep humanism suggests that he would also have rejected the notion of the environment as a determining structure, although for rather different reasons. The problem for Weberian sociology lies not in the essentially 'social' quality of all social interaction but with the idea of structure as a determining influence on the way people behave. In Weberian terms the 'environment' may not be substantially different from society itself. Both society and the environment are composites, forged from much smaller, active units. From a Weberian perspective 'society' and the 'environment' might be considered secondary to the actual interaction occurring between individuals. However, central to the Weberian approach, and to the mainstream sociological tradition, is the notion that human beings are self-conscious actors, aware that their own behaviour can influence wider patterns of social experience. Such a view might seem, superficially at least, very far removed from current political concern over the environment, in which the limits of human ingenuity appear to have been reached. However, as we shall argue, environmental problems have already reached the stage where their solution cannot be left to 'expert' witnesses. Positivist science has itself been the source of many environmental problems, and cannot resolve current human unease over the scale of these problems. There is increasing evidence that environmental policies have failed to address the declining authority of reductionist science. The energies of many groups in civil society have been channelled towards a more affirmative, holistic view of nature.

Finally, for Marx, society is usually conceptualized as a system of social relations. Relations of production, in Marx, remain the basis of material life, and the social relations which govern productive relations are vital to the Marxist perspective. Social production involved relations between individuals and between people and nature. In order to produce material goods people must transform natural resources. This appropriation of nature, however, can only be carried out within a specific social setting, and the Marxist tradition is concerned, centrally, with explaining the context in which the appropriation of nature takes place. Marxism has not been concerned to outline the physical and technical limitations placed on the human capacity for exploitation (of nature and other people) although this question is now receiving more attention than previously (Benton 1993, Dickens 1991, Redclift 1984).

The Marxist tradition involved another element of enormous importance to the current discussion. Marx also asserted that, in their transformation of nature, people are themselves transformed. It is important for our purposes, however, to acknowledge that while Marx considered our relations with the environment as essentially social, he also regarded them as ubiquitous and unchanging, common to each phase of social existence. Hence, for Marx, the relationship between people and nature cannot provide a source of change in society; this is found only in the relationships between groups of people. Such a perspective does not fully acknowledge the role of technology, and its effects on the environment, in altering funda-mentally what it is possible to do with nature. Marxism asserts that economic development under capitalism is about the creation of value as 'resources' are transformed into 'commodities'. But environmental 'laws' like the first and second laws of thermodynamics suggest that matter and energy can be neither created nor destroyed, only converted from one form to another. Any 'value' they acquire in the process needs to be set against these longer-term processes of conversion.

The 'environmental age' in which we live has, as a central concern, to consider whether our ways of exploiting nature are sustainable under any existing political and economic system. The challenge today is to embark on revolutionary changes in the way we organize ourselves to exploit nature, some of which were anticipated by Marx. Notwithstanding the contradictions and confrontations within society, a more immediate problem remains. It is how to limit our getting and spending to activities which do not harm our own ability to reproduce ourselves.

For the 'founding fathers' of modern sociology the natural environ-ment was, on the whole, defined negatively as that which was not 'social'. Possibly as a result of this, it was not until relatively recently that sociologists have turned their attention to the relationship between people and nature. Two major traditions have emerged. Within the structuralist tradition, human ecologists and urban sociologists have tended to view people as a product of their environments, a perspective that is shared by some geographers and anthropologists. The more interpretive tradition, however, has sought to challenge the conceptualization of the environment as a predetermined physical phenomenon which individuals have little capacity to change. Their alternative is to view people as 'goal-oriented', able to define, decipher and explore their physical environments, and to 'construct' a view of nature from within their own consciousness. This perspective rejects environmental determinism, arguing instead for a looser, more interactive, approach to human/environment relations.

Giddens, a leading social theorist, and pre-eminent among modern British sociologists, has developed a theoretical approach which, in effect, marries both the above perspectives, although it has not been developed specifically to meet environmental concerns. His theory of structuration aims to combine structuralist approaches to society, which focus on constraints to human activity, with interpretive approaches, which focus on the intentional acts of human agents. As Dickens (1991) notes, Giddens concentrates his attention on the relations between groups of people rather than on the way in which they exploit nature.

Giddens writes that, 'the reproduction of society is always and everywhere a skilled accomplishment of its members.' It is necessary, he reminds us, to reconcile this fact with the notion that if people do make society they do so under conditions which may not be of their own choosing. Giddens's theory 'expresses the mutual dependence of structure and agency . . . the structural properties of social systems are both the medium and the outcome of the practices that constitute those systems' (Giddens 1979: 69). Giddens argues that structures are enabling as well as constraining factors in the development of individual livelihoods.

The framework which Giddens has called 'structuration' allows us to consider the environment as a stucture which both enables and constrains human agency, while at the same time acknowledging that human agency may change the environment itself. In short, it enables us to take a much broader sociological view of the relationship between society and nature. We also need to explore what this means for the way that the environment is managed, and the social resistance that is frequently mounted to environmental managerialism. These are two sides of the same coin, and both are represented within sociological thinking. Before we look at specific examples of these principles, however, we need to consider why and how social and ecological systems evolve, and the nature of this 'coevolution'.

SOCIETY AND NATURE: AND EXAMPLE
OF THEIR CO-EVOLUTION

Individual societies have contrasting views of 'nature', and of what is 'natural'. In developed, industrial societies the reification of nature has become almost a characteristic. As Goodman and Redclift (1991) have observed,

> Nature has become imbued with so many virtues that the term 'natural' no longer confers unambiguous meaning . . . . We have refashioned nature, in our minds, as well as in test tubes and fields, transforming ecological processes into political axioms . . .
> (Goodman and Redclift 1991: xi)

Differences surrounding 'nature' and what is 'natural' reflect differences between societies. Each society has developed *together with nature* under specific circumstances. At the same time, however, we also need to understand that all development is constrained by nature, by virtue of the fact that nature, in the form of the sun, represents the ultimate source of energy for all human activity.

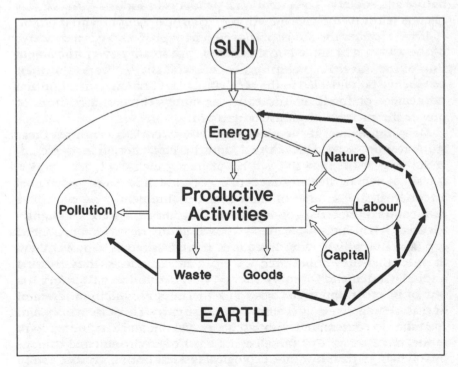

Figure 3.1 Energy and materials in the development process
Source: Adapted from IUCN/UNEP/WWF (1991), p. 76.

We gain a much clearer picture of the 'external' limits of sustain-ability if we locate human activities within the wider domain of physical (and hence 'natural') processes. These more comprehensive 'systems' remain outside the provenance of sociology. We believe they should be brought inside. Figure 3.1 portrays a stylized model of energy and material flows in the development process. The model indicates that development is constrained by energy availability. The ultimate source of energy is the sun, which produces immediately available energy in the form of radiation, wind and the water cycle, and stored energy in the form of plant biomass. This stored energy may either be consumed directly, in the form of food and fuel or, over time, may be concentrated in the form of fossil hydrocarbons. In total, however, there exists a finite amount of incoming solar radiation.

We can trace the energy pathway from the *Sun* to available *Energy* on the *Earth*, where it is utilized by *Nature* and by society in all its various *Productive Activities*. During the production process industry combines *Energy, Nature, Labour* and *Capital* to produce *Goods*, with associated *Waste*. *Waste* material and energy may be recycled or may contribute to *Pollution*, the costs of which are borne by both nature and society. The intentional products of industry, *Goods*, are channelled into the reproduction of *Capital*, *Labour* and *Nature*.

If we consider the industrial development process we can conceive of the way in which the initial harnessing of steam power, the inven-tion of the internal combustion engine, and the realization of their productive potential led to the accelerated exploration, extraction and refinement of fossil hydrocarbons as highly concentrated fuels to power the process of industrial growth.

Over the course of the nineteenth and twentieth centuries these stock resources were developed (and thereby, of course, depleted) at the expense of renewable sources of energy such as wind and water. It was not only particular sources of energy which received preferential attention in the course of industrial development, however. The destination of material goods produced by industry were also tightly focused. Fossil fuels were used to power industries which produced capital goods which, in turn, required more fossil fuels to power them in order that they could produce more capital goods, thus creating a spiralling demand for more energy. In short, industrial society has one of its most important bases in a fuel supply which, in terms of human timespans, is in strictly limited supply. This discussion is of relevance to contemporary social theory. On the one hand our growth model can attempt to internalize the costs of environmental damage by following a path towards 'ecological modernization' (Huber 1982). Alternatively, we may be embarking on a new stage in the develop-ment of post-industrialism, in which we need to elicit the help of non-experts to ensure that we can manage the transition to greater uncertainty or what Beck has called 'the risk society' (Beck 1992).

In an attempt to capture the principles behind this process of resource development Richard Norgaard has coined the term 'coevolution'. This refers to the way in which Western science, resources and the environment have developed as a mutually interactive, coevolving system. Norgaard summarizes his position in the following way:

> Western science facilitated the use of coal and petroleum, but the availability and use of these hydrocarbons, in turn, helped determine the directions and intensity of effort in Western science. The environmental side-effects of fossil hydrocarbon-fuelled agriculture and industry provided a more fertile niche for the environmental sciences. These systems, furthermore, coevolved with the modern social order. The pattern of people living in cities, the organization of people to serve multinational industrial enterprises, the centrality of bureaucratic order,

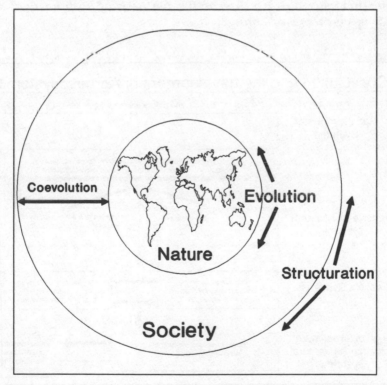

Figure 3.2 Coevolution as a set of equilibrating mechanisms between society and nature
Source: Woodgate (1992).

and the use of Western science for social decision-making, have all coevolved around fossil hydrocarbon-fuelled development. This coevolutionary process resulted in considerable concentration of power and material wealth in modern industrial societies, which they used to force Westernization on others. Simultaneously, non-westernized peoples sought the same power and material wealth through adoption of modern knowledge, social organization, and technology. Correcting the unsustainability of development is not simply a matter of choosing different technologies for intervening in the environment. The mechanisms of perceiving, choosing, and using technologies are embedded in social structures which are themselves products of modern technologies.

(Norgaard 1994: 43–4)

Coevolution, then, can be thought of as a set of equilibrating mechanisms between society and nature (see Figure 3.2). Changes in nature occur through processes of evolution, while changes in society are the result of processes of structuration. In this sense coevolution can be understood *as an interactive synthesis of both natural and social mechanisms of change*.

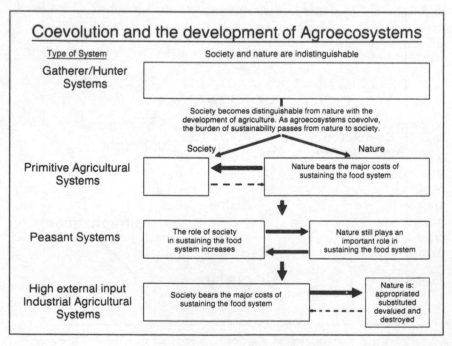

Figure 3.3 Agricultural coevolution
Source: Adapted from Woodgate (1992).

Taking agriculture and the food system as examples, Figure 3.3 illustrates the coevolutionary transition from gatherer/hunter societies to ones which are dependent on industrial agriculture, and require high levels of external energy input (machinery, fuel and agrochemicals) in order to produce their food and raw materials.

In the process of coevolutionary change, society has assumed more and more of the functions traditionally undertaken by nature. In order to illustrate this we can take as an example the differences between small-scale 'primitive' agriculture, and modern, industrialized agriculture. Small-scale agriculture is labour-intensive, polycultural and subsistence-based. Large-scale farming, by comparison, is capital- and energy-intensive, mechanized and monocultural. Large-scale agriculture involves the development of industries to make implements and agrochemical inputs, sophisticated marketing networks and government institutions to generate and disseminate knowledge. These highly complex institutions also regulate markets, absorb risks, limit the distributional impacts of new productive techniques, and seek to control environmental and health-related externalities. Aspects of 'nature' have been refashioned and converted into industrial processes, under scientific control. This is particularly evident in the new biotechnology industries, and in genetic engineering. In the process, it has been argued, problems of the environment require closer management, as society comes to bear the major cost of sustaining the modern food system (Goodman and Redclift 1991).

DEVELOPMENT AND SOCIETY'S EVALUATION OF NATURE

Our willingness to 'manage' nature, and the implications this carries for sustainability, reflects significant shifts in our attitudes towards nature. Some writers draw attention to what they see as a revaluation of nature in the course of economic development. For example, Nash (1973) argues that as societies develop economically they appreciate 'nature' more and 'civilization' less. Figure 3.4 shows what happens, in Nash's view, to the relative valuation of *wild nature* and *civilization* in the course of economic development. The stages through which societies pass can be summarized as follows:

1 Initially, in most developing countries, the marginal value of civilization is much higher than that of wilderness, because wilderness is abundant. This favours the destruction of wilderness, through agricultural enclosure or tree-felling, as the value of natural resources such as forests can be realized through the market. Wilderness is even experienced as a threat to civilization. Nash refers to such societies as 'nature exporting'.

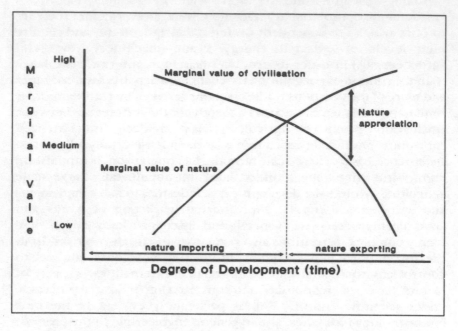

Figure 3.4 Nash's model of the valuation of nature and civilization
Source: Nash (1973), p. 346.

2 With the passage of time 'civilization' takes root and wilderness becomes scarce. Consequently the marginal value of each changes. Societies begin to consider conserving nature as a way of deriving value.

3 After the curves cross in Figure 3.4, society comes to value nature more as wilderness becomes less plentiful. Nature appreciation *increases*. Nash refers to these societies, rather confusingly, as 'nature importing'.

Nash's model serves as an illustration of the culturally bounded way in which both development and the environment are often understood. First, it appears to consider 'development' as a linear process which increases cumulatively over time, rather than a qualitative process which corresponds to changing conceptions of human need. Second, this view is highly ethnocentric. It suggests that we can establish an 'appreciation' of nature, in objective terms, that corresponds with North American or European values. But do all cultures appreciate nature in the same way? And does appreciation of nature increase with economic development? Terms like 'nature' and 'civilization' clearly do not necessarily communicate the same thing in *all* cultures. They are unlikely to mean the same thing to inhabitants of an urban ghetto in New York *and* the Amazon rain forest (Short 1991). They may not even mean the same thing to all the inhabitants of New York (or the rain forest!)

The more we consider the evidence, the less obvious it is that local cultures, and their views of the environment, correspond with those in modern, industrial societies. Norgaard's concept of coevolution helps us to understand the relationship between our own economic development and the view we take of the environment. Unfortunately, environmental management is frequently transposed, uncritically, from the North to the South, in the belief that we can arrive at a 'scientific', objective, account of the environment devoid of cultural content. The supposed objectivity of environmental policy and management receives prolonged and critical attention from other chapters in this volume.

Before we examine other cultural perspectives on the environment, it is worth considering our own. We can begin with the idea of wilderness: one that informed Nash's model in uncomfortable ways. In North America, as in other areas of the globe settled by Europeans, 'wilderness' came to assume importance once it began to disappear. James Stilgoe (1989) has described his understanding of the effect of spreading land husbandry in the United States on the *perception* of wilderness. Wilderness has always existed, but it only assumed value within a Romantic perspective as poets and other writers came to mourn its passing.

In similar vein John Rennie Short (1991) has distinguished two historical perspectives, which different societies have taken towards wilderness. In the first perspective wilderness is overcome, it is 'tamed'. Society thus accords more value to what replaces it, especially sedentary farming. This situation would equate with the 'nature exporting' portion of the curves in Nash's model.

The second perspective Short sees as part of the Romantic Movement in Europe and America during the last two centuries. This views wilderness as a loss that society cannot make up. It represents not the *creation* of the Garden of Eden (human earthly paradise), but human *expulsion* from the Garden of Eden. Both these perspectives are at work in our environmental consciousness in developed countries today. On the one hand, we live in societies whose whole existence has been based on the consumption of more and more natural resources which are assumed to be plentiful; on the other hand, we are haunted by the knowledge that our management of the environment in the past has contributed to many of our current problems. It is not an enormous step to project that anxiety into the future, and into a concern for the welfare of future generations as yet unborn.

The contradictions surrounding our management of nature are also evident in the way we view the 'countryside' which, like wilderness, is closely related to the sense of loss which has accompanied modern industrial 'civilization'. The countryside has assumed 'heritage' status, like cathedrals and the pageantry surrounding the monarchy, by pointing us to our past.

One of the most familiar ways in which we come to know, and to own, the countryside, is through landscape. Landscape and the country-side are social products for two related reasons, which illustrate the way we have internalized our understanding of nature and the environment.

First, they came about as the result of human ingenuity and effort. The countryside is the outcome of human transformations of nature; it does not exist *separately* from human societies. Second, the way we view the countryside and landscape reflects our social values and attitudes at any given time. It varies considerably over time and between cultures. The countryside 'out there' is a countryside of the mind. As both Short (1991) and Rackham (1986) make clear, the social construction of countryside and landscape is a facet of everyday life. It is irrevocably linked to the changes in land uses through which humans have transformed nature.

CONSERVATION IN ECONOMICALLY DEVELOPED COUNTRIES

If the countryside and wilderness are themselves socially constructed, it should come as no surprise that the attempt to manage them, in the name of 'conservation', can best be understood as part of the process of economic development to which industrial (or post-industrial) societies are committed. Industrial societies increasingly view conser-vation as a major priority but – what do they mean by conservation, and what does this tell us about their view of the environment?

In his book *Countryside Conservation*, Green (1981) considers three types of conservation:

> First, to many, conservation is essentially the preservation and pro-tection of those features of the environment though to be of amenity value, such as fine buildings, landscape, wildlife, clean air and water, or even streets free of litter. Some foresee the pollution of air and water reaching levels where our very existence is threatened and they regard the maintenance of environmental quality as much more of a necessity than amenity. . . . [Finally] to others conserva-tion means the planned use of resources to eking out until substi-tutes can be found. This is a much more dynamic idea, embodying change and development, as well as a measure of protection.
>
> (Green 1981: 2)

Green goes on to argue that it is common for people to believe that if resources were to be exploited wisely, the inevitable result would be the maintenance of a desirable environment: that the third type of conservation subsumes the first two. However, such beliefs are founded on a fallacy. Managed environments, in which natural resources are purposefully exploited by human populations, are manifestly different from relatively untouched environments, however wisely or sustainably such environments are managed. Both are 'natural', but in different ways. The exploitation of resources involves trade-offs between productive use, amenity, species conservation, and other environmental goals. Therefore definitions of conservation such as 'the scientific management of natural environments and resources for the purpose of maximizing their aesthetic, educational, recreational and economic benefits to society', although appealing, are in fact misleading, since they ignore the costs of both economic growth and resource conservation.

To provide an example. If a wood-lot is managed primarily for production, then single species stands of trees would be planted in straight lines, with all competing vegetation and potential pests kept to a minimum. The trade-offs associated with maximizing production would include a loss of aesthetic appeal, as well as genetic diversity, and the introduction of toxic chemicals for pest and weed control. Green reflects on this, saying that 'general acceptance of broad definitions of conservation has led [people] to overlook the fact that wise use of one resource does not necessarily ensure good husbandry of another' (Green 1981: 4).

Our view of what conservation constitutes also changes over time. Within recent history there have been quite distinct emphases given to environmental concern in the United Kingdom. Newby (1991) suggests four main phases of environmentalism in the UK:

1 A *preservationist* phase (later nineteenth century) in which growing industrialization gave rise to a disenfranchisement with industrial life and a somewhat romanicized view of the past.
2 A *regulatory* phase (between the First and Second World wars) when environmental concern spread to other social classes, and attention shifted from economic *laissez-faire* to government itself.
3 A *post-materialist* phase (1960s and 1970s) when the social limits to economic growth were admitted, and the quality of the environment was emphasized above material well-being.
4 A *sustainability* phase (1980s and 1990s) in which the emphasis lies with global environmental change, underscored once more by advance in scientific observation. Ecology, rather than amenity, is the focus of public debate, and the consequences of human consumption appear to threaten the very survival of the planet.

These examples are taken from the history of one part of the industrialized North, but they have also constituted an influence on the countries of the South. Transposed to the South, however, these views of conservation may be inappropriate, or even dangerous, since conditions there

63

are radically different. Nevertheless, as the creatures of our societies we often interpret other environments in terms of our own preoccupations. We create ways of managing nature in the image of those we know best, having recourse to the same science and technology and, often, the same apparatus of enforcement and encouragement.

History provides some rich examples of how the idea of conservation, and the image of the environment, has been transferred from one culture to another. Conservation in nineteenth-century Africa, for example, was often carried out for paternalistic reasons that paid scant regard to indigenous experience or cultures. The legacy of these misdirected efforts survives to the present day. Indeed, as McCormack (1989) has made clear:

> the association of wildlife protection with the dictates of colonial administration was largely responsible for the later beliefs of emerging independent African administrations that conservation was a luxury of little relevance to plans for economic development.
>
> (McCormick, 1989: 23)

The colonial imprint had effects that remain important today. Indeed the way that colonial administrations in Africa came to view conservation has influenced the way people in the North perceive Africa's current environmental crisis. It has also influenced the way that people in the South see themselves. Rural populations often possess important knowledge about environmental management, but this has been devalued over time by the penetration of Western 'scientific' knowledge and the search for 'technical fixes' which evade distributional issues entirely. The effects of colonialism in determining the way other countries' natural resources are 'developed' are not merely physical and economic. They are also cultural, influencing the way in which this development is understood.

The discourse surrounding environment and development is not a neutral, convergent discourse, but one reflecting both divergent historical experiences and differing interpretations of those experiences. In the North, countries have adopted their own language to describe environmental problems in the South. This language has become 'globalized', and has served the purposes of the international agencies which have taken the lead in economic development. It frequently conveys the idea that 'environmental problems', and the cost of addressing them, represent a brake on 'development' for countries in the South. Instead of posing questions about the sustainability of 'development', as practised in the highly industrialized countries of the North, the North's environmental concerns have frequently been grafted on to the South's development problems. The result, frequently observed during the 'Earth Summit' in Rio de Janeiro in 1992, is that 'our' environmental problems (climate change and ozone depletion, for example) are caused by 'their' development (industrialization, forest losses and population growth). The alternative vision, which has attracted support from a variety of unlikely quarters, is that of *sustainable development*.

As a recent document expressed it:

> The concept of sustainable development eliminates the need for that ill-defined something called the environment (while it does not eliminate expression of people's cultural and aesthetic responses to nature). For development to be sustainable, it must follow paths that allow for the renewing of renewable resources, whether those resources be local fuelwood or a global climate matched to current agricultural systems. Sustainable development is, by definition, environmentally sound development. . . .
>
> (Holmberg *et al.* 1991: 6)

Other chapters in this collection examine the increasing role of environmental management in containing the problems caused by unsustainable development. The techniques and policy interventions which constitute 'environmental management' in the North are frequently transferred to the countries of the South, as we have seen. What does this imply for sociological analysis? Does our understanding of society and the environment enable us to elaborate an alternative to unsustainable development in which culture and nature evolve together and in harmony? To answer this question, to move as it were towards a 'realist' agenda, we need to make sociology engage with the implications of our relations with nature. We have argued in this chapter that this is both urgent and necessary. Other chapters in this volume suggest ways in which it may be undertaken.

REFERENCES

Beck, U. (1992) *Risk Society*, London: Sage.

Benton, E. (1993) *Natural Relations: Ecology, Animal Rights and Social Justice*, London: Verso.

Dickens, P. (1991) *Society and Nature: Towards a Green Social Theory*, London: Harvester Wheatsheaf.

Durkheim, E. (1982) *The Rules of Sociological Method*, London: Macmillan.

Giddens, A. (1979) *Central Problems in Social Theory*, London: Macmillan.

Goodman, D. and Redclift, M.R. (1991) *Refashioning Nature: Food, Ecology and Culture*, London: Routledge.

Green, B.H. (1981) *Countryside Conservation: The Protection and Management of Amenity Ecosystems*, London: George Allen & Unwin.

Holmberg, J., Bass, S. and Timberlake, L. (1991) *Defending the Future: A Guide to Sustainable Development*, London: IIED/Earthscan.

Huber (1982) *The Lost Innocence of Ecology*, Frankfurt: Fisher.

IUCN/UNEP/WWF (1991) *Caring for the Earth: A Strategy for Sustainable Living*, London: Earthscan.

McCormick, J. (1989) *The Global Environmental Movement*, London: Belhaven.

Nash, R. (1973) *Wilderness and the American Mind*, New Haven: Yale University Press.

Newby, H. (1991) 'One world, two cultures: sociology and the environment', BSA Bulletin, *Network*, vol. 50 (May): 1–8.

Norgaard, R. (1994) *Progress Betrayed: The Demise of Development and a Coevolutionary Revisioning of the Future*, London: Routledge.

Rackham, O. (1986) *The History of the Countryside*, London: Dent.

Redclift, M.R. (1984) *Development and the Environmental Crisis: Red or Green Alternatives?* London: Methuen.

Short, J.R. (1991) *Imagined Country: Society, Culture and Environment*, London: Routledge.

Stilgoe, J.R. (1989) 'Everyday rural landscapes and Thoreau's wild apples', *New England Landscape*, vol. 1, pp. 4–11.

Woodgate, G.R. (1992) 'Sustainability and the fate of the peasantry: the political ecology of livelihood systems in an upland agroecosystem in the Central Highlands of Mexico', Wye College, University of London, unpublished Ph.D. thesis.

4

THE LIMITS TO NEOCLASSICISM
Towards an institutional
environmental economics

Michael Jacobs

INTRODUCTION

Environmental economics is not a new subject; as its historian Juan Martinez-Alier has shown, the relationship between humankind and the natural environment has been an important object of economic enquiry since economics first separated itself from the natural sciences at the start of the Industrial Revolution. Indeed, the basic trajectory of development of the modern discipline was laid down between seventy and a hundred years ago by Jevons, Pigou and other members of the early neoclassical school.[1]

There is no question, however, that it is in the last few years that environmental economics has come of age. Stimulated by the re-emergence of the environment as a political issue and the increasing concern of politicians at the potential cost of protective measures, environmental economics has turned into one of the fastest-growing academic sub-disciplines throughout the industrialized world. Its analyses are already influencing major public decisions, particularly in the United States; large government-funded research projects in the subject have been established in many countries, including the UK; some of its practitioners have even become publicly known names.

In this chapter I want to look at the assumptions and goals of the dominant strand of modern environmental economics: what may be called the neoclassical school. By no means all environmental economists are working within this school: there is extremely interesting and important work being done, for example, in resource accounting, macro-environmental-economic modelling and (even more radically) in ecological and thermodynamic analysis which is only peripherally, if at all, related to the neoclassical framework.[2] But it is fair to say that most environmental economics is being carried out within this framework, and, more importantly, that the most influential work in the subject is of this kind. The neoclassical approach has indeed begun to change the way in which the environment is thought about in public policy-making and in law. Proposals for 'green taxes', the revival of interest in cost–benefit analysis, the use (in the US) of valuation techniques in litigation over environmental losses, and the development of international environmental economic policy by the World Bank and others, all reflect the influence of the neoclassical approach.

It is important at the outset to distinguish the neoclassical approach to environmental economics from another school with which it shares some affinities but which is conceptually separate. This is the 'property rights' or 'public choice theory' school, some of whose practitioners have recently turned their attention to environmental issues.[3] Economists in this school argue that the best way to deal with environmental problems is to assign property rights to environmental assets (water courses, landscapes, endangered species, and so on), thereby creating markets in environmental damage and protection. Only one act, the 'privatization' of the environment, is required of government; the 'optimal' level of environmental protection will then be achieved simply by private bargains between actors in the free market.

This approach shares with the neoclassical school, as we shall show below, a belief that the appropriate level of environmental protection is derived from market transactions; but environmental neoclassicism does not adopt the emphasis on 'minimal government'. Neoclassicists are concerned to create hypothetical, not real, markets (they do not in general advocate privatization of the environment) and are committed to the use of a variety of instruments of government intervention in markets in order to achieve their environmental objectives: taxes, subsidies, tradable permits, etc. It would be fair to say that neoclassicists are therefore market-*oriented*, but they do not advocate (as property rights economists do) *'free'* markets.[4]

In this chapter I shall not be concerned with the property rights approach, which has so far had little influence outside academic circles and right-wing think-tanks. It is the neoclassical school which is setting the agenda. In this context the purpose of this chapter is therefore threefold. First, I want to set out the basic premises and arguments of the neoclassical 'project', premises and arguments which are not always made explicit. Second, I want to demonstrate the limitations of neoclassicism when applied to the environment: of its claim to ethical neutrality, the validity of its approach to environmental valuation, and its ability to model and predict economic behaviour. Third, and too briefly, I want to suggest some potential lines of development for a richer, more interdisciplinary 'socio-' environmental economics. In doing this I shall draw on the 'institutional' or 'evolutionary' school of economic theory, which is also currently undergoing a revival but which has so far paid little attention to environmental matters.

In analysing the neoclassical project, I am glad to acknowledge that I may not be fully or adequately describing the positions of any individual economists. Many neoclassicists will accept – and be working on – some of the criticisms made below. Many will argue that neoclassical techniques, particular monetary valuation of the environment and cost–benefit analysis, should only be used as *contributions* to public decision-making, not as its sole determinants; other criteria and methods should also be employed. Nevertheless I believe that it is useful to set out the basis and 'raw bones' of neoclassicism in order more clearly to understand what is going on when it is being used, however garnished with qualification its presentation in practice. The following description may therefore be taken as the neoclassical 'ideal type', not any particular economist's views.

THE NEOCLASSICAL PROJECT

At heart, the neoclassical approach to environmental economics has one aim: to turn the environment into a commodity which can be analysed just like other commodities. Sociologically it is fairly obvious why this should be done: if the environment is *not* like other commodities and cannot be analysed in the same way economists will clearly have a much smaller role in advising decision-makers on environmental policy. But there are less self-interested reasons as well to do this. Economists see that the environment is frequently undervalued: because it can often be used free of charge it tends to get over-used, and therefore degraded. If only the environment were given its proper value in economic decision-making, the economist reasons, it would be much more highly protected. So far from being the materialist philistines of the Greens' accusation, environmental economists regard themselves as the Earth's true friends, providing arguments for its protection which (unlike those of the emotional Greens) will carry weight in the real world of financial decision-making.[5]

It is important to make this point, for two reasons. First, because in criticizing neoclassicism I do not wish to impugn the motives of those working within its framework. Most environmental economists are genuinely concerned to protect the environment and are engaged in the subject to assist this cause. Second, because neoclassical techniques may well support the protection of the environment; and in these circumstances it is incumbent upon its critics clearly to set out alternatives which are at least as effective. If this is not done, academic criticism may lead to real-world defeat.

Premises

The commodification of the environment in the neoclassical project is based on five premises.

The first is methodological individualism. For neoclassical economists economic activity consists of the behaviour of individuals, and is explained in terms of individual psychology. This means that the environment must be broken down into separate goods and services which are 'consumed' by individuals in the same way that other commodities are consumed. These environmental goods and services provide consumers with satisfaction or utility; we can therefore speak of the 'demand' for it arising from individual preferences at different prices.

The second premise is that of the 'rational economic person'. Individuals behave in a self-interested and consistent manner to maximize their utility.[6] Third, the rational economic person ideally operates in competitive markets where he or she can exercise free choice. Such markets will tend towards an equilibrium state which (other things being equal) will achieve the 'optimal' or most efficient allocation of resources for society: that is, where the marginal cost of the last unit of change just equals the marginal benefit.

Fourth, the neoclassical approach takes preferences and technology as exogenous to the economic system. It therefore treats them as given, rather than as objects of analysis. In combination with the equilibrium assumption of the market this creates an essentially static perspective which ignores the role of time: the world is seen as effectively 'reversible'.

Fifth, neoclassicism tends to regard itself as morally neutral, a 'positive' rather than a normative subject. This is because, its practitioners argue, neoclassical analysis is based on the currently existing preferences of consumers on which it does not presume to make value judgements. It claims merely that, given these preferences, the 'optimal' (but not necessarily the morally best) outcome is the one where benefits exceed costs by the greatest amount. Although neoclassical economists frequently do recommend that society should pursue particular courses of action, they argue that such prescriptions do not constitute normative judgements. Their advice merely represents the policy which best fulfils the preferences of individuals in society.

From these premises, neoclassical environmental economics constructs a powerful case. After an indispensable preliminary exercise, this essentially comes in two stages. The preliminary exercise is to break the environment down into constituent goods and services, the commodities of analysis. These may be the cleanliness of air in such

70

and such a locality, the water quality of this section of river, the preservation of that landscape from this perspective, the numbers of such and such a species, the rise in global temperature over the next thirty years, and so on.

Once this is done, the first stage of the neoclassical project is to give these commodities imputed prices by constructing supply and demand curves. This enables the economist to identify the appropriate (optimal) level of environmental protection for society to adopt. The second stage is to turn these imputed prices into real-life prices by intervening in or creating markets for them. This enables the identified level of environmental protection to be achieved.

The first stage argument: determining the appropriate level of environmental protection

The supply curve for an environmental commodity is based on the costs of the measures required to protect it: not just the direct costs of protection, but the opportunity costs of economic activity forgone, knock-on effects, etc. Calculating these costs becomes more difficult the more indirect they become, but it is widely assumed that the supply side poses few conceptual problems.

Meanwhile the demand curve is estimated by discovering the 'willingness to pay' of the 'consumers' of these environmental commodities. Since most of the products are not actually bought, this is done by imputation, in one of two ways. In the first approach, 'hedonic pricing', consumer preferences are 'revealed' by the real-life market demand for goods which are *associated* with the environmental product in question. Typically, for example, the demand for a quiet neighbourhood will be estimated by comparing house prices in a quiet location with similar houses under a flight path. Or the demand for a beautiful landscape will be revealed by calculating the costs that people incur in travelling to look at it. In either case, isolating the environmental factor is extremely difficult but theoretically possible.[7]

In the second method, 'contingent valuation',[8] individual consumers value the specific environmental commodity in question, but in a *hypothetical* market since there is no real one. This is done by conducting surveys in which a 'representative sample' of people are asked how much they would be willing to pay for the environmental commodity, or (sometimes) how much they would need to be compensated to lose it.

It is recognized that the consumers of the environment are not just those people who use it directly: who actually swim in the river, view the landscape or hunt the species. Environmental consumers are also

71

people who might want to use the environment at some time in the future, who want their children or future generations to have the opportunity to use it, and who want the environment to exist irrespective of anybody's use. These give the environment not just 'use value' but 'option', 'bequest' and 'existence' value also.[9]

From the individual responses to these surveys, the demand curve is built up by aggregating the sample across the total population considered relevant. This achieves the total imputed value of the environmental commodity. In theory, to generate a proper demand curve, a whole range of values should be found, corresponding to different degrees of environmental protection. In practice such a curve is rarely constructed, since the valuation methods can generally only handle one level of environmental protection at a time. At best only a few levels can be handled (making a river fishable, swimmable or drinkable, for example), which makes for a rather limited curve.

But this does not inhibit the neoclassical case. Even if the consumer value of only one level of protection for a particular environmental commodity is found, this is still sufficient to allow a cost–benefit analysis to be conducted in which the costs of that level of protection can be compared with its aggregate consumer value. Such a computation – discounting future costs and benefits at some appropriate rate – will then tell decision-makers whether that level of protection is 'worth it' in economic terms: that is, whether its benefits exceed the costs. In the theoretical case where a proper demand curve with a range of values can be determined, its intersection with the supply curve will indicate the level of protection for that environmental commodity which represents the most efficient allocation of resources in society (the market equilibrium). On the premise that satisfying consumer preferences is what matters, most neoclassicists would argue that this is the level of environmental protection which society therefore should choose.

The second stage argument: achieving the optimal level of environmental protection, optimally

The first stage of the neoclassical project is therefore aimed at *identifying* the appropriate (the most efficient) level of environmental protection. The second stage is concerned with *achieving* this level in the most efficient way. The second stage logically follows from the first, but it can also stand alone. It is important to recognize this. However the desired level of environmental protection is arrived at, the orthodox economist will argue, the neoclassical approach can indicate the least-cost way of achieving it. Because of this, economists working in second stage fields – green taxes, pollution permits and so on – do not feel constrained by criticisms of the first stage. Many of them indeed would reject the first stage argument, asserting that the appropriate level of environmental protection should be determined through scientific, political or other methods with little or no attempt to construct demand curves or to conduct cost–benefit analyses. (Not all environmentalist critics of neoclassical economics have realized this, attacking green taxes and pollution permits on the basis of their rejection of monetary valuation.)

The second stage is much less theoretically complicated than the first. It argues simply that the most efficient way of achieving the desired level of environmental protection is to give environmental costs and benefits 'prices' in the markets where they occur. In this way the value of the environment can be reflected in the real world where decisions are made.

This can be done in essentially two ways. One is to change the prices of *existing* market activities by taxing environmental damage (such as pollution or harmful products) or by subsidizing environmental improvement (such as recycling or pollution abatement). The other way is to *create* markets for environmental goods by issuing permits which are then tradable between firms or consumers. For example, the state could decide to limit pollution in a particular river to a given level by issuing 'pollution permits' allowing pollution to that level but no higher. These permits could then be traded amongst the polluting firms, thus raising the price of pollution.

It is easy to show that, if markets are competitive and economic agents are rational, the use of these kinds of 'financial incentive' instruments[10] will achieve the desired level of environmental protection most efficiently: that is, at least total cost to society. If economic agents have different cost structures for environmental improvement, those for whom reducing environmental damage is cheap will do this rather than paying the tax or buying the permit, while those for whom it is expensive will do the latter. The aggregate expenditure of all firms on environmental improvement will therefore be minimized. This is in contrast to what happens under a regulatory regime, where all firms are required to reduce damage equally, irrespective of the variation in cost, resulting in a greater total expenditure. On these grounds, environmental economists tend to advocate the use of financial instruments for environmental protection wherever possible.

COMMODIFICATION AND CONSUMER PREFERENCES

Most environmental economists are working on either the first or second stages of the neoclassical project. The *Journal of Environmental*

Economics and Management and *Land Economics* are filled with the
results of contingent valuation exercises, discovering the monetary
values of landscapes, wild animals, wetlands, river and air quality and
many other environmental features. Environmental ministries and
bodies throughout the industrialized world are humming to the sound
of economists working out the best form and the consequences of
environmental taxes, charges, subsidies, deposit-refund and bond
schemes, tradable pollution permits and other financial instruments.
For those of us who call ourselves environmental economists this is
welcome employment.

Yet there are questions to be asked about the fundamental basis
of neoclassicism. In this section I want to look at the problems raised
by the attempt to turn the environment into a set of commodities,
an issue which affects in particular the first stage of the neoclassical
project. In the next I shall examine some weaknesses in its understanding of economic behaviour, an issue primarily of concern in the
second stage.

The neoclassical project regards the environment as a set of
commodities, ultimately no different from other goods and services
consumed in the economy. Many environmentalists and Greens object
to this treatment on essentially moral grounds, namely that it devalues
the cultural and spiritual meaning which the environment has for
human society and ignores altogether the rights of other species.

This is an important objection, to which we shall return. However,
there is a rather more immediate one which needs to be answered
first. This is that the environment in general clearly *isn't* a commodity
which is traded and individually consumed like ordinary 'produced'
goods and services. It is frequently not 'owned' in a straightforward
way, it is often consumed free of charge, and it is generally public
not private in character. Allocation of environmental goods is generally
under public regulation rather than that of the market. The environment is indeed (in most cases) a classic example of a public good. That
is, it is a good whose 'consumption' is indivisible: it is impossible to
separate off one person's from another's. It is elementary economic
theory that the value (total benefit) of public goods cannot be derived
from individual market preference, since no individual will be prepared
to buy a good on which others can then 'ride free'. Economists do
not expect defence, policing, lighthouses and other public goods to
be valued and therefore determined in this way: they accept that
political decisions will be made on appropriate levels and (compulsory)
taxes levied to pay for them. One might have expected that the
environment would be treated in the same manner.

The important point here, therefore, is that when neoclassical
economists treat the environment as a commodity they are doing

something quite different from what they do when they treat toothpaste or tomatoes or other produced and traded goods and services as commodities. In these cases economists are analysing what is actually happening in the world, how real transactions occur between consumers and producers in real commodities. They therefore have some grounds for arguing that they are engaged in a positive science, examining and making policy recommendations on the basis of people's actual preferences.

In the case of the environment, however, they are (generally) not doing this. They are analysing what *might* happen *if* the environment were a set of commodities and consumers and producers had to make market choices. They are therefore not examining and making policy recommendations on the basis of people's actual preferences but on the basis of what those preferences might be if circumstances were different. This can hardly be described as a positive scientific activity: indeed it is at first sight a rather strange activity altogether. Why should economists analyse a thing as if it were something else?

Optimality, ethical neutrality and institutions of social choice

The neoclassical economist has two answers. The first is that although the environment is not a commodity in real life it should be. This is because – subject to certain assumptions – markets are the most efficient way to allocate resources, and ultimately the environment is a resource which provides human society with benefits just like other resources. If the environment is *not* treated like a commodity, society will allocate it 'sub-optimally' – protecting it too little (less than people are prepared to pay for its benefits) or too much (more than they are prepared to pay). In general the environment exhibits 'market failure'. The neoclassical economist therefore seeks to create a replacement market to make good the damage, with the aim of allocating resources to where they will bring most benefit.

But of course there isn't market *failure* because there wasn't a market in the first place. (The language betrays the implicit assumption that the existence of a market is in some way 'natural'.) And in these circumstances it is not clear that we should automatically wish to create one. For while allocating resources 'optimally' in the economist's terms (that is, where the marginal cost of the last 'bit' of environmental protection equals its marginal benefit, aggregated across all those participating in the market) is certainly *one* way of conceiving the 'most benefit' to society, it is not the only way. It is in fact a utilitarian criterion: it is concerned only with individual preferences, and it measures only totals for all individuals not distributions between them.

Given the choice, there are a number of other criteria we might wish to consider in allocating resources. We might be concerned about the *distribution* of resources. We might tolerate some inefficiency in total allocation to ensure a more egalitarian sharing-out – for example, insisting that everyone should have access to the same quality of drinking water. It is evident from the hypothetical market designed by environmental economists that individual preferences are a function of income – the 'willingness to pay' criterion cannot be divorced from ability to pay, which leads inevitably to inegalitarian outcomes.

Similarly, we might be concerned to respect what we feel are people's *rights* over different aspects of the environment – for example, not being prepared to see a rain forest destroyed (however valuable its timber or land) if this destroyed the culture and livelihoods of its indigenous inhabitants. Again, the hypothetical markets neoclassicists create show little regard for rights: asking people what they are willing to pay to preserve a particularly loved part of the countryside in the face of a threat from development has often appeared to respondents as more like a protection racket than an attempt to discover 'preferences'.[11]

We might also wish to consider the impact of current decisions on *future generations* – for example, we might want to reduce carbon dioxide emissions, even at some present cost, to prevent future global warming. Again there is little room for future people in the hypothetical markets of contingent valuation. This is curious in a way, since the whole basis of environmental economics is the problem of 'externalities'; and a great deal of 'green' concern arises from the long-term nature of environment degradation. This is inevitable, so long as this analysis is based on current markets (real or hypothetical), since future people are not present in current markets and exercise no bargaining power within them. Discovering the option, bequest and existence values of the environment expressed by current consumers does not address this problem, since these values, while certainly giving weight to the future, do not express the interests of future people. They express the interests of the present generation in those people, which is not the same thing.

The interests of *other species*, of course, are similarly excluded. (It is a well-known result of environmental economics that the extinction of slow-growing species such as whales can be 'optimal').[12]

All these factors might be used instead of optimality (or in some way combined with it) as criteria for social choice. To privilege optimality is therefore to make a straightforwardly normative value judgement that this is in some way the best or most appropriate criterion. This is, of course, a perfectly legitimate thing for economists to do, but in doing so no claims should be made to ethical neutrality. Nor should the economist expect society to regard his or her judgement as having any greater authority than that of anyone else.

The neoclassical economist is ready for this argument. Who is this 'we', he or she asks, this 'society'? Who is making these alternative judgements? When markets generate social choices they are doing so on the basis of individual people's preferences. It is not the economist who decides, but everyone, expressed through self-determined choices. By contrast this 'we' being invoked sounds suspiciously like a cover for dictatorship: not everyone's preferences but the author's own.

When the commodity in question is toothpaste or tomatoes, where markets do exist and people do express preferences in them, this argument might carry some weight (though it is not conclusive).[13] But in the case of the environment it falls down at the first hurdle because *there are no individually expressed preferences*. This is the very problem that we started out with: it is no use invoking the primacy of individual preferences to prove the primacy of individual preferences. The importance of optimality is simply a value judgement.

The issue at stake here is not really the normative nature of neoclassical economics. It is the design of the institutions through which social choices are made and enforced. One possibility is the market. In markets individuals do exercise self-determined choices, but these are particular types of choices. Because of the circumstances which markets present, people generally have to make decisions on their own, unaware of how these decisions will add together with those of others to generate overall social outcomes. Even if there is some awareness, individuals will rarely have an incentive to act 'co-operatively' because of the problem of the prisoner's dilemma: if no-one else does it, it has no effect (and the individual loses out). So individuals generally make decisions in their own self-interest.

The other option is for social choices to be made through some political or collectively organized process (such as a public inquiry), and for the decisions to be legally enforced. (That is, the environment might be treated as if it actually were a public good.) Here, of course, the process and outcomes may be of many different kinds; but they can, if so wished, reflect the moral criteria (or non-self-interest) mentioned above, and democratic voting procedures allow individual views (including self-interest) to count. Of course, some people may not feel 'represented' in such institutions; but this is also true of markets, where only those whose preferences happen to coincide with a sufficient number of other people's to generate the result they desire are 'represented'. In a market, of course, those with more bargaining power are more likely to be represented; in a democratic procedure (and not all social institutions are democratic, of course) each individual counts the same.

We shall return to the issue of social institutions below. For now, it is sufficient to note how far the neoclassical project has strayed from its supposed purpose of positive explanation. These are intensely political questions; and while no-one should deny the economist the right to participate in them, the political ramifications should not be hidden.

Environmental values

Neoclassical economists do not have to rely on political opinions to justify their approach. There is a second answer to the question why economists analyse the environment as if it were a purchased commodity. This is that it is as a purchased commodity that people in fact behave towards it. To the neoclassicist, methodological individualism and the rational economic person are not merely heuristic assumptions; this is actually how people are believed to think and act. With regard to the environment they are essentially seen as frustrated by the non-existence of a market, trying to behave in a 'normal', market-oriented way, consuming the environment up to the point that the cost of doing so exceeds its benefits; but without a proper market in which to do so. The attempt to elicit consumer preferences for environmental goods and services, particularly through the creation of hypothetical markets, is therefore designed to reveal what is implicitly going on anyway but which is hidden because the absence of a market prevents preferences being expressed. The economist in these circumstances is seen as a sort of structuralist literary critic, digging down beneath the surface of 'real' life to see what is really going on underneath.

The problem here is that this is a hypothesis, not an obvious fact. Therefore what we might have expected environmental economists to do is to test it, not use it as an assumption. They might have designed their contingent valuation exercises to find out whether this is, in fact, how people implicitly behave towards the environment. In actuality, the methods are specifically constructed to ensure that people can only participate if they take this stance. For example, respondents are not allowed to talk to one another (even when the exercises are in a laboratory) despite this being the obvious way in which a sense of collective value for a public good might be derived, and the prisoner's dilemma overcome.[14] And if respondents simply refuse to play the game, unwilling to make monetary 'bids' for environmental goods in the way expected of them, they are excluded from the results as 'don't knows'.

The remarkable thing about contingent valuation surveys is that, though they are expressly designed to reveal individual market preferences, in fact they often seem to do just the opposite. In many cases the results of such surveys can just as fairly be interpreted to show that people do *not* value the evironment as a marketable commodity in this way.

First, large numbers of people do refuse to participate in the surveys. Indeed researchers have been forced to acknowledge a category not just of 'refusal' but of 'protest', where the reason for non-response has explicitly been given. In several cases, particularly in valuation exercises for environmental goods such as famous landscapes and endangered species, up to 50 per cent of respondents have simply refused to participate, arguing that the exercises are an inappropriate method of expressing their environmental values.[15] Such behaviour would appear to be strong evidence that the neoclassical approach does not in fact reflect how some people, at least, think about the environment.

Second, even where respondents have participated, the meanings of the values generated are by no means clear. The researchers have found that willingness to pay 'bids' vary greatly according to the amount of information respondents are given about the environmental feature in question, the 'starting point' for bidding suggested and the 'payment vehicle' through which they are hypothetically exercising their preference. Practitioners describe these variations as 'biases', and have been assiduous in trying to eliminate them.[16] But this simply begs the question, since a value can only be biased if there is a true value from which it diverges. But it is this very true value which the neoclassical economist is trying to discover by these methods. Rather than trying to eliminate the bias, it seems at least as legitimate to ask whether the respondents are actually expressing determinate values at all. Asked 'How much would you be willing to pay to preserve the Wyoming bighorn sheep?', it seems quite reasonable to answer 'don't know' or 'it depends . . .'

Third, and connected to this, there is a real problem over aggregation of the values derived in surveys. Each exercise is undertaken separately with respondents asked their willingness to pay or to accept compensation for a particular aspect of the environment. But this does not mean that these values can then be added up to get a total for all the different aspects together. If asked in a *single* survey to pay for all of them, respondents would surely be very unlikely to come up with a figure which was simply the aggregate of all their bids for the individual aspects (or even nearly so) – if only because, given the kinds of results achieved in the separate surveys, this would lead to people spending an unrealistically large proportion of their total income on environmental protection. (A qualification should be entered: namely, that some degree of aggregation is plausible for aspects of the environment which are directly experienced, such as local peace and quiet; it is for those aspects, such as preservation of wildlife and countryside, which have strong elements of 'existence value', that aggregation poses problems.)

But if this is the case, then it is not clear what the original individual bids mean. Isolated from other spending decisions, it is not evident that they can be interpreted as market preferences. An alternative interpretation would be that such bids are essentially statements about people's willingness (and ability) to pay for general environmental protection; even as described in some places, for 'moral satisfaction'.[17] But of course this is not what the neoclassical economist needs them to be.

Fourth, there is a consistent problem in contingent valuation exercises about the difference between willingness to pay (WTP) and willingness to accept compensation (WTA). In theory these values should be approximately the same. In practice, WTA is always higher, sometimes by up to a factor of five.[18] The principle reason for this is that people commonly value losses from their current endowment of environmental assets or consumption levels more highly than they value apparently equivalent gains. This is in itself a very important finding, which has far-reaching policy implications – not least it suggests that policies based on willingness to pay consistently underestimate environmental losses and therefore allow too much environmental degradation.[19] But there may be another factor involved as well in contingent valuation exercises where 'existence values' are being measured. It may be that whereas willingness to pay is limited by ability to pay, WTA values effectively express a different, non-monetary kind of valuation altogether. Asked, 'What would it take to compensate you for the extinction of the blue whale?', many people would surely be inclined to answer 'nothing'; they might well find the question absurd, if not offensive. But how would this be expressed in a contingent valuation exercise? The answer 'nothing' would tend to *reduce* the value generated. So high compensation bids are given, and WTA exceeds WTP.[20] Neoclassical economists now tend to dismiss WTA for goods which are not already and directly enjoyed by respondents; but it may be that it is the only way non-monetary, non-preference based valuations can be expressed within the confines of the exercise.

Even if this interpretation of contingent valuation exercises is only partially correct, it throws doubt on the neoclassical assumption.

Methodological individualism and the rational economic person may misrepresent the way in which people value the environment. People may not (simply or only) have consumer preferences expressible in money terms as if the environment were a commodity purchased in a market. In many cases they may think and behave towards it in another way altogether.

Mark Sagoff argues that the problem here is that neoclassical economics makes a 'category mistake': it is not *preferences* which people have for the environment, but *attitudes*. The environment belongs in the sphere not of monetary but of moral valuation: people choose what they believe to be *right*, as 'citizens', rather than what is in their *interests*, as 'consumers'. Protecting the environment is thus more like conducting a jury trial than buying a consumer durable: whether or not it should be done is a matter for moral and social debate, not a calculation of costs and benefits.[21]

Under this argument, environmental economics has no place in determining the goals of environmental policy at all. The first stage of the neoclassical argument would simply be eliminated. Environmental goals would be set by political debate; the costs of protection would need to be calculated, but there would be no role at all for the monetary valuation of its benefits.

In this way, Sagoff argues, neoclassical economists have simply misunderstood what is going on when people 'value the environment'. But as with the neoclassical case, the arguments here can very easily elide empirical assertion and moral judgement. Sagoff and other critics do not only believe that the attitudinal, 'citizen-based' conception is how people *actually* value the environment. They believe that this is how people *should* value it.

And it is for this reason that the arguments over contingent valuation and monetary valuation of the environment become so heated. For what this debate is really about is not psychological interpretation. It is not even about the outcome of environmental policy. As neoclassical economists are always at pains to point out, monetary valuation almost always bolsters the case for protection, since it raises the hurdle over which environmentally damaging projects must leap in any cost–benefit analysis. The debate is really about the social and moral structure of society. Should more and more decisions, even ones about public goods, be left to markets, devoid of wider moral consideration? Or should some be retained within the sphere of political institutions, subject to the interests of the 'public good'?

The critical element to note here is how the *premises*, and not just the empirical results, of the competing theories contribute to the ideological debate. For what seems evident about attitudes towards the environment, not least from contingent valuation surveys, is that many people, perhaps most people, are not sure themselves how they think about the environment. They know they value it in some way; for different aspects of the environment they may have various views on how far it should be protected. But how far they regard it as a consumer good towards which they can express market-type preferences, and how far as an object of public, moral decision, we may suspect that most people are not sure. And in these circumstances the language in which theorists express the issues may become very influential.

If neoclassical economists come to dominate the debate on environmental policy, people may begin to think in neoclassical terms. They may start to regard the environment simply as a bundle of goods and services, in which there ought to be markets to determine the appropriate levels of protection. They may begin to feel that it doesn't matter if poorer people do not have equal access to clean air and water, or at least no more than it matters already that they do not have equal access to other commodities. They may start to think that the interests of future generations and other species should only be taken into account in so far as individuals are willing to pay for them in market transactions.

In this way the language and theory of neoclassical economics may actually come to influence the behaviour it is intended to analyse. There is already some evidence to show that if people understand a theory they are more likely to behave in the ways it predicts: in laboratory exercises on public goods economists tend to free-ride more than other people, apparently because they know that this is how the theory says people behave.[22] Similarly, it may be that if the neoclassical approach is widely accepted, people may get used to valuing the environment as a consumer good, and begin to do so. In this way neoclassical environmental economics may in a sense become self-proving.

INSTITUTIONAL BEHAVIOUR

In the second stage of its argument, the ideal type of neoclassical environmental economics relies on a standard model of economic behaviour in which firms and consumers maximize their profit in conditions approximating to perfect competition. It is on the basis of this model that neoclassical economists are often to be found arguing that financial incentive instruments are the most efficient means of achieving environmental goals.

But this is not a model which conforms to the real world. As elsewhere, it ignores the ways in which lack of information, time constraints and institutions and culture affect behaviour. These factors can severely undermine the value of the neoclassical prescription.

For example, it is widely argued by orthodox environmental economists that the best (most efficient) way to achieve a reduction in pollution is to raise its price by taxing (or, more accurately, charging for) polluting discharges. But this will only be true where firms behave 'rationally' in maximizing their financial profit. In some cases, the evidence is that firms don't do this. For example, raising sewage discharge fees in Britain by up to 400 per cent over a five-year period failed to induce any change in firms' behaviour, despite the ready availability of technologies which could reduce discharges within payback periods of less than a year.[23] This was because even the raised fees were a low percentage of firms' total costs, so management time was spent on achieving other savings; and because the firms did not have (or seek) information on the available technologies. Indeed surveys revealed that most firms did not understand how the charging system worked. Legal regulations on discharge levels and permitted technologies, which firms *did* understand, would almost certainly have achieved a reduction in pollution at lower overall cost. (Moreover, a successful charging regime would have required at least as much administrative cost in providing information to firms as a regulatory one.)

Similar evidence can be found in the field of energy. Again, neoclassicists have tended to argue that energy or carbon taxes are the most efficient means of raising energy efficiency. This would undoubtedly be true if the energy market, its consumers and producers, behaved rationally according to the neoclassical premise. But evidence suggests that they do not. In some cases this is a problem of lack of information (for example, about energy efficiency measures). Partly it is because the management costs of undertaking investments are too high. Partly it is because of the 'landlord–tenant' problem, in which those responsible for capital investment do not pay operating costs. Partly it is a function of the structure of the energy market, in which the supply of energy efficiency services (such as insulation) is fragmented in comparison with the supply of energy itself. Because firms and households typically have higher discount rates than energy utilities, from a social point of view (a social discount rate which gave the future welfare of the community a higher value) there is under-investment in energy efficiency services and over-investment in energy supply. Partly it is simply because of inertia: psychological factors inhibit firms and households from making investments (for example, in energy efficiency lighting) even when payback periods are short.

For all these reasons, it can be shown that energy taxation on its own is actually *not* the least-cost method of achieving consumption reductions: a whole variety of regulatory measures, such as energy efficiency standards and least-cost requirements imposed on energy utilities, are likely to cost society less overall.[24]

Of course, many neoclassicists would recognize these problems and are able to (and do) introduce information asymmetries and other structural features into their analyses. In so far as this is done, neoclassical practice moves away from its ideal type towards a more rounded model of the world; such a convergence between neoclassical and more structural or 'institutional' economics is obviously welcome. Different prescriptions for economic policy, of course, should follow.

TOWARDS AN INSTITUTIONAL
ENVIRONMENTAL ECONOMICS

I have argued that neoclassical environmental economics is an inappropriate or too limited a tool with which to understand many environmental issues. I would like to propose that an *institutional* approach (which might also be called *evolutionary* or *socio-economic*) will provide us with a richer, more explanatory framework.

Institutional economics can trace a tradition from Thorstein Veblen, Joseph Schumpeter and J.K. Galbraith through to modern post-Keynesians. A brief summary of the institutional approach might be as follows. Institutional economists reject the methodological individualism that underpins the neoclassical analysis. They argue that economic behaviour is culturally determined, and that institutions in society (such as governments, regulations and property rights) are not 'market imperfections' but the very structures which allow markets to operate. They argue that economic tastes and preferences are endogenous to the economic system and should be objects of economic analysis. They see the static, equilibrium-based approach of neoclassicism as abstract theorizing, and stress the irreversible dynamic nature of real ('evolutionary') economic behaviour. Institutionalists acknowledge the value-basis of theory, accepting and pursuing the ethical presumptions of their arguments.[25]

By and large institutionalists have not so far addressed environmental questions, being concerned principally with conventional macroeconomic and industrial issues. But the general approach seems particularly appropriate in the environmental field. A number of research projects for an institutionalist environmental economics suggest themselves.

First, an institutionalist approach would want to ask, what is actually going on when people 'value the environment'? Neoclassicists say that people are expressing consumer preferences; their Sagoffian opponents that it is moral judgements which are being exercised. The one view claims that economists alone can prescribe environmental policy; the other that economics has no role to play at all.

But surely both these positions are too simple. It seems much more likely that people have *both* economistic preferences for *and* moral and cultural attitudes towards the environment. It is quite possible for people to want to protect different aspects of the environment (or not) for a whole variety of reasons belonging in different 'categories'. It is plain that, as well as being a moral concern, some features of the environment are in some senses a purchased commodity. We do have to pay for environmental protection, and sometimes these payments are indeed quite individualized (it is quite plausible, for example, to argue that house prices reflect local air quality, views and noise; here the environment *is* individually purchased). The truth is surely that environmental valuation is an extremely complex matter, analysis of which requires a multiplicity of approaches.

Institutionalists would therefore wish to conduct empirical research which teased out how people were actually thinking and behaving towards different aspects of the environment: their work would be informed by psychology and sociology as much as by economics. They might conduct surveys, like contingent valuationists, but designing them in such a way that alternative *types of valuation*, and not simply different levels of monetary value, could be revealed. They might convene 'focus groups', as in modern market research, to bring out through discussion how different people feel about different types of environmental decision. Indeed such work is now beginning to be done – emerging out of the contingent valuation approach as its practitioners recognize the limitations of the neoclassical ideal type and extend their work into a more institutional framework.[26]

In such projects one of the questions would no doubt be what role the language and theory of economics itself has in influencing people's stance. Methodological individualism and rational economic person might then be one way of expressing the *conclusions* of empirical research in some instances; they would certainly not be its *assumptions*. But equally, the bald distinction between 'consumer' and 'citizen' behaviours as made by Sagoff and others would need deeper examination. Can motivation be broken down into two simple categories in this way? If so, when and how? Institutionalists would want to find out.

The second element of the institutionalist project would follow from the first. This would be the attempt to develop social institutions which are able to reflect how people think and behave towards the environment, so that public decisions are more able to represent a democratic choice. Here again neither of the options on offer appears to be adequate. Neoclassicists offer cost–benefit analysis. But this has all the problems we have already suggested: a utilitarianism which discounts the interests of poorer and future people, which ignores rights and cannot cope with non-monetary valuation. On the other side Sagoff offers a 'black box' of public decision-making, displaying a naïve faith in the processes by which elected politicians come to policy conclusions.

Institutionalists would want to examine how in practice, and how ideally, political institutions do and should work to make decisions on the environment. In this they would wish to collaborate with researchers in political science and sociology. They would want to ask how information is gathered, interpreted and presented; what role different groups within the community play in the process; who has rights to influence and make decisions (different scales of government, local communities, affected individuals, businesses, landowners) and to what degree; how different arguments (including those of neoclassicists and Sagoffians) are represented and reconciled. In devising appropriate institutions the economists' own normative view would inevitably be expressed; but this would be open and explicit, not hidden and denied as is too often the case in the neoclassical project.

Third, the institutionalist project would address the question of economic behaviour discussed in the second stage of the neoclassical project. How do firms and households actually behave in real markets, as opposed to the idealized ones of neoclassical theory? Here again the assumptions of methodological individualism and rational economic person would be abandoned as inadequate tools to consider real behaviour. Institutionalists would recognize in environmental behaviour, as in all other aspects of economic behaviour, that economic actors are constrained and shaped by the social, cultural, psychological and ethical context in which they act. The institutional model would consequently not assume away, but would actively study, institutional and cultural factors such as information lack, management costs, attitudes towards discount rates, and structural inertia. It would examine markets not as 'natural' phenomena which somehow exist and which sometimes 'fail', but as dynamic, social, evolutionary institutions which are created and shaped by social, economic and political forces.

Informed as it would have to be by other disciplines, such as industrial sociology, organizational studies, and psychology, such an institutional environmental economics would offer a richer, more dynamic view of the world, and one more likely to offer prescriptions which will contribute to the protection of the environment. True, they will be less mathematical: demand curves will be less precise, econometric models possibly absent altogether. Fewer generalizations will be made; there will be more consideration of specific circumstances. But this is surely as it should be in the social sciences; the Newtonian paradigm – economics as physics envy – was surely never appropriate in the analysis of human behaviour.

Fourth, and as part of this, the institutionalist project would seek to understand how tastes and demands come to be formed. Neoclassical environmental economics has barely addressed this question, seeing tastes as exogenous to the economy and therefore outside its remit. But it is surely implausible to argue that economic activity does not help to form demand as well as satisfy it. In the environmental field, institutionalists would wish to examine how people come to have particular attitudes and values – not just towards the environment but towards those goods and services which lead to its degradation. How does the typical demand of Western societies for consumer goods arise? What is the role, not just of advertising (the obvious target) but more obviously of infrastructure provision, in affecting this demand? It seems evident that town planning has been radically transformed by the development of the car; but that in turn such planning has required most people to have a car. Might there be alternative patterns of economic development with different consumption behaviours and different levels of environmental degradation?

Such an analysis, drawing on sociology, anthropology, social history and social psychology, would acknowledge the changing patterns of tastes and demand over time. As societies develop socially and economically, tastes and wants change. An institutional environmental economics would need to study this process in order properly to understand environmental behaviour, seeing the economy as a dynamic and historical system rather than one in reversible equilibrium.

The fifth element in the institutionalist project would follow from this. It would seek to understand and measure the 'standard of living'. Orthodox economics of all kinds (not just neoclassical) considers the standard of living purely in terms of disposable income. Yet it is surely appropriate to consider other factors too in judging economic welfare. Considerable contributions are evidently made to welfare by publicly provided and non-marketed goods and services, of which the environment is a prominent example – in terms of health impacts, amenity and life support functions. Institutionalists would wish to find new ways of measuring these 'quality of life' elements, so that public policy can be informed by them. Since the standard of living is such a critical part of political debate, and orthodox economics is its dominant language, this may be seen as an important political project as well as an academic one.

CONCLUSION

As neoclassicists will always (and rightly) remark, it is much easier to deconstruct orthodoxy than to replace it with a more convincing approach. As they have also observed in the environmental field, neoclassical approaches can in many circumstances assist the cause of environmental protection. The development of a convincing institutionalist framework of analysis and policy prescription is therefore crucial and urgent.

Institutionalists must also recognize that the neoclassical tradition has done a great deal to raise the profile of economics within environmental policy, and to some extent vice versa. The integration of the economic and environmental field is a critical intellectual and policy advance, and the neoclassical school has greatly assisted this process. Many neoclassicists would themselves recognize the problems associated with the ideal type of their project, and some are working to overcome them. In this context the criticisms raised here should not be regarded as setting up a simple bipolar antagonism between neoclassical and institutionalist approaches. In so complex and little understood a field a variety of methodologies is required: there are surely valid contributions to understanding to be made by different schools in different circumstances. In a world of methodological pluralism, we may perhaps lay down two conditions. First, that economics allows into its analysis contributions from other disciplines, resisting its frequent tendency to intellectual hegemony; and second, that attention remains focused on the real world of actual environmental behaviours and urgent environmental problems, not on the formal abstractions to which economics is prone.

NOTES

1   J. Martinez-Alier, *Ecological Economics: Energy, Environment and Society* (Oxford: Basil Blackwell, 1987).
2   For interesting examples see (among many others) R. Constanza (ed.), *Ecological Economics: The Science and Management of Sustainability* (New York: Columbia Press, 1991); P. Ekins and M. Max-Neef (eds), *Real Life Economics* (London: Routledge, 1992); H. Daly and J. Cobb, *For the Common Good* (Boston: Beacon Press, 1989); Y.J. Amad, S. El Serafy and E. Lutz (eds), *Environmental Accounting for Sustainable Development* (Washington, DC: World Bank, 1989); R. Hueting, P. Bosch and B. de Boer, *Methodology for the Calculation of Sustainable National*

Income (Voorburg, the Netherlands: Netherlands Central Bureau of Statistics, 1991); R. Repetto, M. Wells, C. Beer and F. Rossini, *Wasting Assets: Natural Resources in the National Income Accounts* (Washington, DC: World Resources Institute, 1989); T. Barker and R. Lewney, 'A green scenario for the UK economy', in T. Barker (ed) *Green Futures for Economic Growth* (Cambridge: Cambridge Econometrics, 1991); R. Norgaard 'Environmental economics: an evolutionary critique and a plea for pluralism', *Journal of Environmental Economics and Management* 12(4), 1985.

3 See in particular T. Anderson and D. Leal, *Free Market Environmentalism* (San Francisco: Pacific Research Institute for Public Policy, 1991).

4 The confusion that has arisen in some quarters in understanding the different schools is partly due to a loose use of the term 'market'. In terms of policy neoclassicists are *not* 'free' marketeers; they advocate extensive government intervention in markets through environmental taxes and subsidies, etc. Property rights economists do not propose such measures: if the environment is privately owned, they are unnecessary. In terms of the distinction I make below between the 'first stage' of environmental economic policy (identifying the appropriate level of environmental protection) and the 'second stage' (achieving it), the property rights school elides the two together. There is only one stage: the appropriate level of environmental protection is whatever the free market – achieved by privatization – throws up.

5 See for example, D. Pearce (ed.), *Blueprint 2*, 'Introduction' (London: Earthscan, 1991).

6 This is generally taken to mean that they maximize their financial profit, but if challenged will usually be adjusted so that non-financially-self-interested behaviour can be tautologically described as utility-maximizing. For critiques see K. Lux and M. Lutz, *Humanistic Economics* (New York: Bootstrap Press, 1988), and Etzioni, *The Moral Dimension: Toward a New Economics* (New York: The Free Press, 1988).

7 For a survey of hedonic pricing methods, see P-O. Johansson, *The Economic Theory and Measurement of Environmental Benefits* (Cambridge: Cambridge University Press, 1987).

8 Strictly speaking, contingent valuation is not the only form of hypothetical preference, but the other method, 'stated preference', has been rarely used and is little advocated. For an application and discussion, see MVA Consultancy, Institute for Transport Studies (University of Leeds) and Transport Studies Unit (University of Oxford), *The Value of Travel Time Savings* (Newbury: Policy Journals, 1987). For a survey of the contingent valuation method see R. Mitchell and R. Carson, *Using Surveys to Value Public Goods: The Contingent Valuation Method* (Washington, DC: Resources for the Future, 1989).

9 There is also a concept of 'quasi-option value', which is the value of preserving options for future use given some expectation of the growth of knowledge (for example, the medicinal value of tropical rain-forest species).

10 Sometimes called market mechanisms because they use the price mechanism which operates in markets. However, if this term implies that regulations, which are the alternative to market mechanisms, do not operate (or even abolish) markets, it is unhelpful. Moreover, some have

89

taken the use of this term to mean that such mechanisms are a 'free market' solution to environmental problems, which as interventions in the market they are evidently not. The ideological connotations of these terms are best avoided.

11 P. Hopkinson, C. Nash and N. Sheehy, *How Much Do People Value the Environment: The Development of a Method to Identify How People Conceptualise and Value the Costs and Benefits of New Road Schemes* (Institute of Transport Studies and Department of Psychology, University of Leeds, 1990).
12 C. Clark, *Mathematical Bioeconomics* (New York: John Wiley, 1976).
13 It is not conclusive because even though there *is* a market for ordinary commodities, this does not mean that there *ought* to be, or more precisely, that the outcomes of the market are the morally best ones. To make this leap is to elide the economist's role of observation of the world with moral judgement upon it.
14 This is called in the literature 'strategic bias'. See R. Cummings, D. Brookshire and W. Schulze (eds), *Valuing Environmental Goods: An Assessment of the Contingent Valuation Method* (Totowa: N.J.: Rowman & Allanheld, 1986), pp. 21-6, 207.
15 See, for example, R. Rowe and G. Chestnut, *The Value of Visibility: Economic Theory and Applications for Air Pollution Control* (Cambridge, Mass.: Abt, 1982), pp. 80-1.
16 See Mitchell and Carson, *Using Surveys to Value Public Goods*. These biases are discussed in M. Jacobs, *The Green Economy* (London: Pluto Press, 1991).
17 D. Kahneman and J. Knetsch., 'Valuing public goods: the purchase of moral satisfaction', *Journal of Environmental Economics and Management* 22, 1992, pp. 57-70.
18 Mitchell and Carson, *Using Surveys To Value Public Goods*.
19 J. Knetsch, 'Environmental policy implications of disparities between willingnesss to pay and compensation demanded measures of values', *Journal of Environmental Economics and Management* 18, pp. 227-37.
20 R. Gregory, 'Interpreting measures of economic loss: evidence of contingent valuation studies', *Journal of Environmental Economics and Management* 12(4), 1986, pp. 325-37.
21 M. Sagoff, *The Economy of the Earth* (Cambridge: Cambridge University Press, 1988).
22 G. Marwell and R. Ames, 'Economists free ride: does anyone else?', *Journal of Public Economics* 15, 1981, pp. 295-310.
23 M. Pearson and S. Smith, *Taxation and Environmental Policy: Some Initial Evidence* (London: Institute for Fiscal Studies, 1990).
24 See T. Jackson and M. Jacobs, 'Carbon taxes and the assumptions of environmental economics', in T. Barker (ed.), *Green Futures for Economic Growth* (Cambridge: Cambridge Econometrics, 1991).
25 A useful modern account is Geoff Hodgson's *Economics and Institutions* (Cambridge: Polity, 1988). Amitai Etzioni's *The Moral Dimension*, op. cit., is a valuable attempt to define a socio-economics. The *Journal of Economic Issues* is the principal forum, while the European Association of Evolutionary Political Economy and the Association for the Advancement of Socio-Economics have both been recently formed to develop a broadly institutional approach.

25 See, for example, G. Peterson 'New horizons in economic valuation: integrating economics and psychology', in M. Lockwood and T. DeLacy (eds), *Valuing Natural Areas: Applications and Problems of the Contingent Valuation Method* (Albury, NSW, Australia: The Johnstone Centre of Parks, Recreation and Heritage, 1992), along with the chapters in the same volume by Lockwood, Blamey and Cameron.

5

RUNNING OUT OF TIME
Global crisis and human engagement
Barbara Adam

Environmental crises are gaining prominence on sociopolitical and academic agendas. There is widespread consensus about what constitutes the crises but little agreement about their causes and potential cures: what is seen as a solution from one perspective is often considered a source of the problem from another. Focus on aspects that are normally ignored allows us to sidestep those hardening oppositions and provides a new access to environmental issues and their analyses. While the spatial dimension has been brought to the fore in a number of disciplines, the temporal equivalent has stayed implicit. Explication of what remains invisible in conventional analyses is used to shed new light on the familiar, agreed upon and disputed.

INTRODUCTION

There is considerable consensus that we are facing an environmental crisis. Its increasing seriousness has put environmental issues high on the sociopolitical and scientific agendas. Widely accepted as contributors to the crisis are the pollution of land, air and water and prominent among these are the depletion of the ozone layer, global warming and the disposal and rendering harmless of waste. No such agreement, however, has been reached over interpretations of the degree of the problems, their causes or their potential solutions;

> There are underlying disagreements over how problems are defined, their degree of seriousness, who is responsible for solving them, and how amenable they are to solution. These disagreements run deep; they are based on different moral principles, different values, different assumptions about how the world operates, and they are found not only at the international level, where cultural diversity is to be expected, but at all

92

levels, within a single society or organisation, and within the actions and policies of a single corporate group.

(Milton 1991: 4)

Often, some group's answers are another's causes and, 'what is rational from one perspective is deeply irrational from another' (Porritt 1984: 15). The goals of industrially sustained growth and ecologically sustainable growth, for example, are difficult to reconcile. To propose a strategy of 'best available technique not entailing excessive costs' (BATNEEC) is incompatible with the proposal for a new ecological ethic which insists on a shift from industrial product to ecological process, from planned obsolescence to durability, from short-term to long-term solutions and from understanding the measure of gross national product (GNP) as progress to viewing it as problem and pollution indicator.

Given this pervasive social debate, it is not surprising that sociologists are increasingly involved in the study and exploration not of environmental matters *per se* but of their attendant social relations, perspectives and actions. And yet there is unease about this social science approach to environmental research. How can sociologists make sense of the different definitions, approaches and proposals for solutions when their discipline has been established on the irreducible distinction between nature and culture, the natural and the symbolic environment, evolution and history, when the environment as a subject matter so clearly falls outside their traditional bounded domain? One strategy has been to stay clear of the substantive issues and to focus instead on environmentalism, the rise of green issues on the political agenda, on assumptions underpinning deep and shallow ecology, as well as the social construction of those scientific 'facts' (Cotgrove 1982, Dobson 1990). Environmental issues *per se* could then be safely left to the 'scientists most suited' to their analysis and explication. At the same time, however, there is dissatisfaction with this social science approach (Newby 1991). That is to say, the present strategy seems somewhat misplaced when the hazards threaten our health and our survival – as members of an academic discipline, of families and societies, of a species and as living organisms on this planet – and when our present actions affect future generations thousands of years hence. Marx's comment that it is not enough merely to study society but to change it never seemed more pertinent. Concern with *substantive issues* and with *engagement* rather than detached explication, however, demand some far-reaching changes at the very heart of sociology. Such a stance necessitates first that we open the disciplinary boundaries and extend our focus to the living, physical and artefactual environment. It requires, further, that we transcend

93

the dualisms which constitute the very foundations of social science: nature–nurture, fact–value, matter–mind, observer–observed, objectivity–subjectivity, to name just some of the most central of those antinomies. It means that we need to redefine the *scientific* component of our discipline and incorporate in that reassessment an engagement with the subject matter, an acceptance of multiple, irreconcilable perspectives, and a recognition that we are creating a fundamentally unknowable future. Lastly, engagement suggests a standpoint that unifies the investigator, the victim, the 'activist' and the bystander.

Environmental processes have no regard for socially constituted boundaries: they span the globe and affect both living and inorganic matter. They impact on people across nations, regions, institutions, and disciplines. Their study therefore has to be adequate and appropriate to that particular characteristic. This poses problems for empirical research which can only deal with single, specific, observable phenomena in succession and then reconstitute the complexity by assembling the isolated components. Where empirical research is bound to the local and specific, theory can reach out. It can transcend the empirical restrictions and explore the complex, the uncertain and the invisible. It can aspire an understanding of wholeness and knowledge of the realm beyond sense data. It is here, at the theoretical level, that we first have to make sense of the uncertainties created by our actions, the incompatible findings and the hazards that have to be faced by us now and by future generations for hundreds, even thousands of years to come. To aid this process of re-vision it is helpful to take a phenomenological attitude: to focus on the taken-for-granted, that which is normally ignored.

In this chapter I am therefore not concerned with the history of environmentalism or the social construction of scientific data. Instead I want to identify some of the unattended time dimensions of the environment–industrialism interpenetration and explore their implications for social theory. I focus on a number of distinguishing temporal features of artefacts, machines and environmental processes and make explicit the complexity of times that permeates specific aspects of contemporary environmental crises. My express purpose is to span the gaps between personal, professional and environmental concerns in a way that suggests at each of these levels active participation in the creation of the future.

OF ORGANIC AND ARTEFACTUAL TIME[1]

Time in the natural environment is characterized by rhythmic variation, synchronization and an all-embracing, complex web of interconnections.

Linear sequences take place but these are part of a wider network of cycles as well as finely tuned and synchronized temporal relations where ultimately everything connects to everything else: the structure of an ecological system is temporal and its parts resonate with the whole and vice versa. Rhythmicity, therefore, forms nature's silent pulse. All organisms, from single cells to ecosystems, display interdependent rhythmic behaviour. Some of this rhythmicity constitutes the organism's unique identity, some relates to its life cycle, some binds the organism to the rhythms of the universe, and some functions as a physiological clock by which living beings 'tell' cosmic time. The natural environment is thus a temporal realm of orchestrated rhythms of varying speeds and intensities as well as temporally constituted uniqueness. It is also a world of organisms with the capacity for remembering and anticipating, of beings that time their actions, synchronize their interactions and reckon time. The very essence of life, furthermore, is growth and evolution. In organic processes, therefore, the entropic principle changes its direction from decay, uniformity and heat-death to growth, variation, and life. This involves the creation and regeneration of time. The use and depletion of time are counterbalanced by its generation and replenishment: decay is compensated by repair through healing and by 'super repair' through the birth–death cycle. Moreover, natural processes vary with context. This means that general principles find unique expression: the rhythmically changing constellations of the stars never repeat themselves in exactly the same way. Springtime, the period when a large proportion of land-based nature comes to life, is incomparably different from wintertime when so much lies dormant. Lastly, a vast range of time-spans coexist simultaneously. These extend from the imperceptibly fast to the unimaginably slow, covering processes that last from nanoseconds to millenia.

The artefactual world of human culture differs significantly from many of these temporal characteristics of living beings. Though often conceived as copies of nature, artefacts do not remain embedded within the give and take, the transience of ecological interconnectedness and exchange. They are created apart, frozen for contemplation, fixed in their uniqueness. They take on a material existence with a difference: externalized, abstracted, bounded and isolated. They are created as islands of permanence in a sea of creative change. Their existence constitutes a finite time, encased in things and isolated from the processes of life and ecological interconnections. Consequently, their temporality is governed by entropy not development and growth. The emphasis is no longer on process but product. Except for their most recent expressions, those cultural products encapsulate the aim of longevity. Paintings and books, cathedrals and nuclear power, all are efforts to overcome individual death, all are moves towards immortality (Becker 1973). Through artefacts human beings are able to gain knowledge beyond their personal worlds and extend themselves into the past and future. To create in isolated and unchanging form what is interconnected, moving and changing facilitates reflection, contemplation and study. It advances the growth of knowledge and provides the potential to have a relationship to that which is other than Self. Culture, from the Latin *colere*, to tend or to till, emphasizes both the relationship to nature and the separation and distinctiveness between culture and nature.

This cluster of time-related characteristics of artefacts – externalization, disembedding, detemporalization and emphasis on longevity and precision – is implicated in the construction of tools, the transformation of raw materials, and the creation of cities. As such it is central to the human impact on nature. Artefacts created with the aid of science are therefore not unique in this sense. Rather, they have further intensified those discontinuities between human production and the temporal ongoings of living processes and with it they have vastly extended the impact of human activity on the environment. Like other products of culture, steam engines, cars and nuclear plants are not temporally embedded within the ecological give and take of their environments. Unlike other artefacts, however, these products of science are rhythmically organized and finely tuned within each isolated system. This makes machines both similar and different from living systems, which are not only temporally organized but also temporally embedded in a rhythmically organized energy exchange where the discarded energy of one system constitutes the source of energy for another. This machine–organism distinction further contains a different meaning of efficiency. Life tends to depend on high energy exchange within systems as well as between systems and their environments, whilst the efficiency of a machine relates to entropy and low energy exchange; minimum entropy means maximum efficiency. In other words, there has to be just enough energy difference to make the machine work. The aim is to achieve a differential that is as close as possible to non-dissipation. An efficient car, therefore, uses the minimum of petrol for the maximum of distance travelled. An efficient forest, in contrast, uses a maximum of carbon dioxide (waste energy of others) and transforms it into a maximum of oxygen (central life source for others). Entropy is implicated in yet another important organism–machine distinction. Machines start to deteriorate as soon as they are constructed and the sources to counteract this process are always located outside the system rather than being generated from within: machines are not designed for death and regeneration. The success of living systems, in contrast, depends on their mortality and on the transience of their internal subsystems. 'The constant deterioration of the molecular and cellular components', Morin (1974: 563) suggests, 'is the weakness which gives living beings the advantage over machines.' Living being, he argues, is constituted on the basis of disorder whilst the design of machines is premised on 'the high reliability of their component parts'. These distinctions, however, should not be conceived as simple dichotomies: temporal organization within systems or within and between systems, minimum or maximum entropy production, entropy or growth, order from order or order from disorder. The differences are far more multifaceted and mutually influencing: disorder tends to result from the creation of order. There is no growth without entropy production. Long-term stability is invariably based on short-term impermanence and uncertainty. Complexity and implication, rather than opposites, are thus necessary guides for understanding the role of human culture and contemporary science in the present environmental crisis.

A second feature, exacerbated by the scientific way of life, is the relation to the environment as external and 'other', and the treatment of nature as an inanimate source of exploitable resources. This Western relationship to nature as 'other' has a number of very deep rooted sources in pre-scientific praxis (in the Marxian sense of unification of thought and practice). Central among these are the linear-perspective vision and the creation of a clock time. Both are powerful externalizers. Both separate subject from object. Both are devices that distance us from experience. Both facilitate mathematical description, quantification, and standardization. The linear-perspective vision of reality, a fifteenth-century technique to represent three-dimensional space on a two-dimensional plane, is a conceptual as well as an artistic method that enables us to employ a window ethic. It allows us to abstract observer from observed, and the part from the whole. Before its inception, the world and our place within it looked and felt differently: the person was always integral to the system of observation, which meant that knowledge was fundamentally relative. With the linear perspective the body was moved from the centre to the outside, from the midst to outside the scheme of things. This entailed a transformation of participants into spectators, of active agents into passive observers (Romanyshyn 1989). An analogous process occurred with the relationship to time when clock time became abstracted from the complexity of times that pervade our existence, when time became disembedded from the temporality, tempo, timing, rhythmicity, chronology and historicity of being and separated from the past, present, and the future. With the creation of an independent machine time, the multiplicity of times was transformed into an objectified, measurable quantity that could be used as a social tool for orientation, co-ordination, synchronization, and regulation. Temporalities thus mediated could then be translated into an object, a material commodity to be used, allocated, sold and controlled. While the linear perspective translated depth levels into spatial distance, clock time represented the passage of time as distance travelled in space, or as measurement of length. Emphasis on visual space, distance and detachment corresponded to the visual representation, externalization and decontextualization of time. The linear-perspective and clock-time vision found their coherent and full expression in the scientific world view. In other words, as conceptual tools they encompass a number of principles that can be observed in scientific theories and designs: emphasis on abstraction, separation and otherness, elimination of surrounding context, and the allied pursuit of permanence and timeless truth. 'But the divorce of inner from outer, above from below', as Roszak (1992: 6) points out, 'could never be more than a temporary expedient, a way of getting on with fact gathering.' Today, this particular cluster of characteristics is being questioned: it is considered inappropriate for the description and analysis of our globally networked reality. The subject–subject relation, for example, is no longer the prerogative of the human sciences: biologists and philosophers extend that relation to other living species, contemporary physicists to the physical environment both on earth and in the cosmos. It is now the turn of sociologists to extend their subject–subject relations to animals, plants and inorganic matter and to the reembedding of cultural activities in the temporal ongoings of nature. It is time for engagement with the environment.

RUNNING OUT OF TIME

'We are running out – not of resources but time.'
'All study, planning, negotiation and implementation of agreed action takes time.'
'We could not foresee the speed of depletion.'
'The rapid time-scale of deterioration means that we have to cut our planned time-scale of action by half.'
'We have to bring the cut-off date forward by five years to 1995.'
(BBC 1992)

Although these statements, made in a recent BBC radio programme on environmental issues, relate to the thinning of the ozone layer, they resonate equally with a number of politically less attractive environmental threats such as global warming, deforestation, acid rain, and the depletion of topsoil. Moreover, they are no less relevant to matters of waste: the export of Northern societies' waste to the poor countries of the South, the dumping of toxic waste in the oceans, and the hazardous accumulation of radioactive waste. Global problems need globally co-ordinated action, which is difficult to achieve and extremely slow. Such global action involves people from different nations getting together to discuss, explore and plan, feed back to their countries – nationally and locally – allow for discussion, consultation, and more research, construct scenarios, explore feasibility and costs, suggest new solutions and compromises. After that, a next round of talks begins at the global level. Invariably, the time-frame of the perceived danger is out of sync with the time frame for action and all too often the exigency of the crisis is traded against political and economic interests, national pride and established habits. This has the effect of widening the gaps between the time-scale of the problem, the time-span of concern and the horizon for action. Not surprisingly, a sense of urgency and a fear of being too late, of time running out for effective action, pervades some of the debates. While all types of pollution share the difficulty of out-of-sync time-frames, none are as easy to identify and solve as the depletion of ozone. None can be shown to have a single, direct, and provable relation between input and ouput, cause and effect, between a human product, an environ mental crisis and human health. None require the phasing out of merely one product or action. This makes the thinning of the ozone layer unique among environmental effects and so attractive for political action: the proven cause presents a challenging but *manageable* problem and, most importantly, it can be tackled without a change of values, economic practices or political structures. It needs no more than the 'speedy' phasing out of chlorofluorocarbons (CFCs) and and the invention of an alternative substance to replace the many functions performed by CFCs. Not surprisingly, therefore, this particular environmental danger has sufficient global support to be carried through to what is at present envisaged to be a successful conclusion. By June 1990 a proposal to halt the production of CFCs by the year 2000 was agreed by 93 nations (Brown 1991: 19). By April 1992 this cut-off date has been brought forward to 1995 (BBC 1992).

Despite the uniqueness of ozone depletion as an environmental threat, the history of CFCs shares many characteristics with other originally benign human inventions whose unforeseen and unpredict-able consequences have turned into environmental hazards; all were considered technological triumphs, contributors to progress. Indeed, immense care and consideration had gone into their design. Only with hindsight could their dangers be recognized. The emergence of such unexpected problems is to be found equally in the history of steam, petrol and jet engines, the harnessing of electricity, gas and oil, and of many other scientific innovations. CFCs are synthetic gases which have a wide range of applications in aerosols, fridges, air-conditioning, insulation and fire extinguishers. They have, as Yearley (1991: 13–14) points out, an extraordinary range of benign qualities. Developed early this century, they were hailed to be non-inflammable, non-poisonous, and non-reactive with other substances. Certain CFCs can be liquefied under pressure and when the pressure is released they evaporate rapidly. Most importantly, they do not break down easily. They are designed for longevity, which makes them very cost-effective. Their creators could not have envisaged the havoc this invention would wreak in the upper stratosphere of our earth. It was outside the predictive capacity of the designers, producers and promoters of this product to foresee that its most positive qualities – unreactiveness and longevity – would cause the dangerous thinning of the ozone layer with its attendant malignant effects on earth. It took highly sophisticated scientific analyses, unavailable at the time of invention, to identify the indirect, non-visible links between this particular cause-and-effect relation, between CFCs and their depleting action on the protective layer in the upper stratosphere which selectively screens out high-energy ultraviolet radiation from the sun and thus enabled the evolution of our present form of life on this earth. At the beginning of the twentieth century this particular future was unknowable in the same way as the effects of other human developments were unpredictable at the time of their proud inception. An unknown and unpredictable future thus constitutes an integral part of the history of novel scientific inventions.

The story of CFCs is characterized by a further time characteristic: it is replete with time lags. It is marked by time lags not merely between cause and effect, or between invention and recognition of the problem, but also between the identification of the problem, its multinational acceptance and the global agreement to take action. It entails time lags between the will to action and its collective execution, between actions and effects, between human corrective change and environmental recuperation. Despite corrective action, for example, there may be long periods of worsening effects before any sign of improvement can be registered. Such multiple time lags constitute an integral part of most environmental problems. Time lags, gaps, and latency periods often mean that the links between causes and effects have become invisible, that their relationship is no longer amenable to scientific certainty and verification. In the specific case of CFCs the problem has been known since the early 1970s and extensive warnings have been voiced since that period. However, it was not until 1985, when a major reduction in ozone was reported and the increase in skin cancers could no longer be ignored, that large numbers of countries agreed to phase out CFCs by the year 2000. Even then, however, there was still great reluctance and a general clamour for 'proof' (Churchill 1991: 157–60, Yearley 1991: 17–21). This insistence on certainty and proof for situations characterized by uncertainty, unpredictability, multiple time lags and out-of-sync time-frames is central to much of the political complacency about environmental problems: nothing is done until the connection is proved. More research rather than action tends to be the response from business and politicians. Proof, in the conventional sense of empirical science, however, is impossible to achieve when there is no directly observable link between input and output, when the relation is not 1:1 but 1:many, when we are dealing not with static phenomena but with continuously changing situations and parameters, when the reactions are latent and invisible for long periods of time and when the effects are manifested not in the location of perpetration of the act but in neighbouring countries or even on the other side of the globe. The idea of proof takes on a new meaning when it is based not on the verification of observable 'facts' but on confirmation of speculative theories.

Global warming is another case in point where those temporal features re-emerge. Here too, there is little consensus about the severity of the problem, its time-scale and the most cost-effective solutions. While governments seek proof as a pre-condition to instigating 'controlling' action, the scientific discourse is filled with expressions of uncertainty: 'Sooner or later it will affect vegetation and agriculture'; 'cereal growing areas of the USA may become too hot'; 'it will probably affect weather systems'; 'the climate may become more extreme'; 'it would begin to melt the ice caps and sea levels would rise in step'; 'it is difficult to establish how rapidly the effect is working' (Yearley 1991: 17–18). Moreover, there is a multitude of competing estimates and calculations based on different combinations of variables, most of which are characterized in turn by time lags, latency periods and poorly understood relationships. The past in these cases is not a reliable predictor of the future and even where there is measured trust in past-based knowledge, scientists realize that the time-scales of their investigations outstrip the time available for action, that time will run out before they can provide reliable answers. As Pugh (1989: 29, in Yearley 1991: 18) explains about the spectre of rising sea-levels, 'analyses of sea level trends need at least 20 years of measurements. There are no short cuts, no ways of speeding up the steady accretion of data.' Here too, traditional approaches to knowledge seem to constitute a poor basis for analyses of out-of-sync time-frames and fundamental uncertainty; materialist assumptions and past-based knowledge suit technological solutions and traditional

101

politico-economic action but are ill matched for the task of dealing with complex, multiple interaction networks characterized by unpredictable latency periods, invisibility, and seeming non-causal connections. The actions planned or already taking place are not the only choices available: the options were much wider than those shorter-term solutions which now dominate the debates (Brown 1991: 18–20, Churchill 1991: 162, Yearley 1991: 19–21). A global programme of reforestation, for example, would be equally if not more effective. With respect to this option, however, the opposite is taking place: deforestation on an unprecedented scale. This loss of plant cover not only reduces the earth's capacity to transform carbon dioxide into oxygen and to cool the atmosphere in the hottest regions of the globe, but also the cutting down and burning of forests adds to the carbon dioxide emissions of industrial societies. Thus, deforestation in tropical regions contributes between 20–30 per cent to global warming and some deforestation of the Northern territory of Canada, covering only 0.017 per cent of the earth's land-surface, contributes 2 per cent to the problem (Postel and Ryan 1991: 80).

The depletion of plant cover in general, and of forests in temperate climates in particular, is further hastened by acid rain which is formed from sulphur dioxide, another by-product of the burning of fossil fuels. This in turn affects carbon dioxide emission and thus enhances global warming. There is an additional connection therefore between global warming, the death of trees and lakes, corrosion of buildings, and health hazards. Once more, the relation between action and outcome is not direct. The effects of acid-forming actions often manifest themselves in different times and places, thus making proof, predictability, accountability and willingness to act extremely difficult to achieve. The export of this hazard by industrial countries makes acid rain an international environmental problem; its link with the greenhouse effect turns it into a global one. Forests, their development, growth and regeneration show up once more the discrepancy of time-frames between evolving nature and the impact of industrial capitalism. There is no overlap, no touching even of the thousands of years it takes for a forest ecosystem to establish itself and the short time it takes to destroy it. Equally out of sync are the time-frames of damage and regeneration. More important still, this discrepancy is not exhausted by the number of years it takes to grow individual trees. If the soil's fertility has been damaged or eroded, the regeneration of that alone takes hundreds of years, if it is possible at all. The time-scale to re-establish the complex net of interdependencies of the above- and below-ground communities of plants and animals which constitute forests as integrated living systems can take as much as thousands of years. This problem of regeneration is intimately tied to two factors: fragmentation and monocultures. It is not just the felling of trees *per se* but the way these biosystems are depleted which seems to be of crucial importance since this affects the forests' ability to adapt and regenerate. Forests, like other bio-systems, are characterized by strong, self-referencing feedback loops (*Journal of Sustainable Forestry* 1991). These processes appear to aid the system's stability and its resistance to disturbances typical of their environment such as floods, fires and storms. Adaptability of this kind seems to result not simply from the complex interaction between physical, biological and even human social forces over time but rather, as Maser's (1991) research shows, the forest's ability to 'migrate as interactive aboveground-belowground communities of symbiotic plants and animals'. The fragmentation of habitat, caused by deforestation, therefore has the most devastating effect on the forests' capacity to adjust to environmental changes: vast areas of fragmented plant cover may not be able to adapt to the global warming brought about by industrial countries of the northern hemisphere. A key characteristic of the scientific, industrial way of life thus constitutes a pervasive threat at the level of ecological existence.

A second important feature of forests as adaptable systems is their multiplicity of plant and animal species. Imposed monocultures, in contrast, appear to be ill-suited to adapt to environmental stress. The pine forests of Germany and Scandinavia dying from acid rain are an important case in point. Thus, long-term sustainability may well crucially depend on both diversity and connectivity. To counter-balance industrial carbon dioxide emissions, it may not be enough, therefore, to plan reforestation with monocultures on a massive scale. As Maser stresses,

> we must look beyond the sustainability of forests as isolated entities in time and space to the long-term sustainability of forests as contextual components of landscapes, which must be designed to be adaptable to changing environmental conditions over time.
>
> (Maser 1991: 55)

Such an approach, however, runs counter to the central features of the scientific technology of abstraction, isolation, singularity and decontextualization. It is not clear, moreover, to what extent these insights have filtered through to the majority of governments and their relevant policy-makers, but it seems vital that they do since such knowledge is central to the maintenance and expansion of plant cover not destined for furniture and beefburger production (Brown 1991, Ekins 1992, Palmer 1992). Once the climate can no longer be taken for granted as a constant, our actions and policies have to be adapted

103

accordingly. More than before, there will be a pressing need to shift from practices that foster short-term profit to those that aid the long-term ability of total systems to adapt to the industrially-induced changes to the environment. There is hope, but not much trust, that the economically inspired options based on BATNEEC, the best available techniques not entailing excessive costs and BPEO, the best practicable environmental option (Churchill 1991, Holliday 1992, Warren 1991), will be ousted by REV, a more *radical environmental vision*. A central obstacle to such a vision, however, is the economic, commodified relationship to time.

As I have shown elsewhere (Adam 1990: 112–19), in societies with commodified time speed becomes an economic value: the faster goods move through the economy the better; it increases profit and shows up positively in a country's GNP. A very different picture emerges when speed is linked not to profit but to energy use. When time is associated with energy we become aware that the faster something moves or functions, the higher tends to be its use of resources: it transforms speed from something to be aspired to into a liability. Tied to energy, speed does not mean profit. Rather, it constitutes a deficit on the balance sheet: not gross national product but gross national pollution.[2] Space, too, is implicated in that energy–time fusion. For the military, for example, increased speed means an increased land area for manoeuvre. As Renner (1991: 3) demonstrates, 'a World War II fighter plane required a manoeuvring radius of about 9 kilometres, compared with 75 kilometres today and a projected 150–185 kilometres for the next generation of jets'. The need for manoeuvring space on land has increased equally: from 1 square kilometre per 100,000 soldiers in ancient times to 248 square kilometres during the First World War and 55,000 square kilometres for the same number of soldiers at the 1978 NATO manoeuvre in West Germany (Renner 1991: 134). A third relation exists between the speed of action and the scale of effect. Conventional and nuclear wars in their preparation and execution are cases in point where the speed of action stands in a direct relation to the spatial and temporal scales of their effects: the faster the action, the larger the effect. An equally devastating, but inverse, relationship seems to pertain between the human capacity to have an impact on the environment on the one hand and the loss of control over the effects on the other: the larger the impact the less control we seem to have over the consequences. This relationship is nowhere more evident than in the history of nuclear power.

With the invention of nuclear power, humans have created an invisible, alternative reality that reveals its secrets (but by no means all of them) exclusively to those capable of creating and maintaining it: the nuclear physicists. Other scientists may measure visible effects such as levels of pollution, contamination and mutations in living organisms. For the rest of humanity the hazardous world of radio-activity remains invisible until it materializes into symptoms of contamination. Even then, the linkages can be denied on the basis of insufficient proof, in the absence of scientific means by which we can demonstrate the direct link between input and output, between nuclear 'plants', nuclear accidents and, for example, the dramatic rise in childhood leukaemias. While causal relationships cannot be ruled out, as the Black Report (1984) argued in the case of Sellafield, the con-nections could not be 'proven'. Invisibility, latency periods and the relativity of place give an illusion of safety and security (Macgill 1987). We are surrounded by the effects of nuclear power even though we can't see, hear, touch, taste or smell them. From a materialist perspec-tive, therefore, the inherent, persistent, all-pervasive dangers of nuclear power are both difficult to conceive and to quantify.

Risk assessment and evaluation of liability are cases in point. The potential enormity of the time-scale and extent of a nuclear disaster are such that this particular risk falls outside the capacity of conven-tional systems of insurance. As Beck points out,

> [with] nuclear power plants [we] have suspended the principle of insurance not only in the economic but also in the medical, psychological, cultural and religious sense. *The residual risk society has become an uninsured society* with protection paradoxically diminishing as the danger grows.
>
> (Beck 1992: 101)

Compensation for radiation damage inflicted by the military, for example, is no more than a distant dream for East European citizens and an unlikely occurrence in the USA. A mere US$200 million of government trust funds were set up by the US government for specific groups of victims, and the military are shielded from additional liability to claims which could amount to billions of dollars (Renner 1991: 146). The situation is no better in the public sector. Due to the lack of operational experience and the extent of potential claims, as May (1989: 13) shows, the US government was 'forced to legislate for a limited liability' of $560 million set up in a federal fund. 'The nuclear industry was to have no residual liability; the public would have no common-law right to bring any claim against the builders or operators of any accidentally damaged plant.' In Britain, the potential cost of both accidents and the disposal of radioactive waste made nuclear power unviable for privatization. This exemption from conventional insurance gives us an indication of the discontinuity of nuclear technology with earlier scientific inventions central to industrial societies.

The longevity of nuclear materials as sources of energy and as waste has created environmental dangers for which there are no historical precedents. With nuclear power more than any other environmental threat, therefore, the past holds no clues for our actions in the present and affords no predictive power over the future. Present knowledge about the potential effects of nuclear waste disposal polices is based on guess-work and statistical probabilities. Moreover, we have no materials equal, let alone superior, to the life-span of the waste material we need to make safe. To encase high-level radioactive matter in glass (or medium-level waste in concrete) and to bury this deep under ground or at sea means our actions today create substantial health hazards and general environmental disasters for successive generations in the not-too-distant future. 'Solutions' with such a high degree of mismatch between time-frames may let the people responsible for those actions sleep easy in the knowledge that they, their children, and may be their grandchildren are possibly 'safe'; but nobody can know with certainty today how many generations may be secure on the basis of the durability of glass (and concrete). Nobody can be sure about the exact manifestations and the precise effects of the disintegration of those 'protective' materials on future generations at the receiving end of 'safety' measures implemented during the second half of the twentieth century. Notions of safety, control, predictability and proof are clearly incompatible with a threat where the potential of the impact stands in an inverse relation to the capacity for control. With nuclear power, science loses not only its 'sharp edges of clarity' (Macgill 1987: 13) but its privileged status of provider of truths and certainties. Moreover the gap is ever-widening between the time-scale of the problems and the time-span of concern. We are not, as many suggest we must be, 'mindful of the consequences our decisions will have on the generations of the future and their ability to respond to the conditions that our decisions [and actions] will create' (Maser 1991: 56).

The longevity of products and our inability to re-embed our creations in the temporal structure of the environment is one important dimension of the contemporary problem of waste: nuclear, chemical, metal, or plastic. The other relates to the sheer quantity of consumer products and the brevity of their use. Up until a decade ago the fear was that the world would run out of non-renewable resources. Today there is a recognition that the environmental damage will become intolerable long before those natural resources are exhausted: we will be running out of time, not resources.

The danger of such high levels of consumption lies less in running out of resources, as was commonly argued in the seventies, than the continuing damage that their extraction and processing impose on the environment . . . rising levels of carbon dioxide in the atmosphere make it unlikely the world will run out of oil before the environmental cost of its use – in form of global warming – becomes prohibitive.

(Young 1991: 41)

There is a need to extend planning and audits to include the materials' entire history, to chart their paths from extraction to the rubbish dump, from the 'cradle to the grave', to adopt a life-cycle rather than a point-in-time perspective. This is necessary because it is at those extreme ends – extraction and obsolescence – that most of the irreparable damage is inflicted on the environment. At present, those earliest stages in the conversion from embedded raw material to product and the final change from product to waste rarely feature in the economic calculations about the costs of materials and products, despite the fact that environmentalists have presented strong cases in favour of the life-cycle approach. Adoption of the wider time-frame demonstrates further that waste occurs not only at the end of the path, that instead a large proportion of waste is generated early in the production process. Mining, for example, generally requires the removal of more than double the quantity of unwanted material in relation to the resource being sought. In the USA mining amounts to approx 6–7 times the amount of municipal solid waste produced (Young 1991: 41–2). Solutions to the problem of waste, therefore, must not be singular but multiple: to calculate costs from 'cradle to grave', to shift emphasis from product and profit to process and cost, to make things last longer by reusing and repairing them, and to re-introduce the value of durability. Such an approach, however, fits uneasily with the present economic ethic of maximum sales and the tight association between speed and profit. The dictum of maximum sales means the briefest possible use and the cheapest possible construction. It means advertising to create a need for consumption. It means replacement not repair, planned obsolescence not reuse. It means emphasis on export and long haulage of goods across the globe. While *sustained* development is an economic concept based on an expectancy of economic growth the ecological concept of *sustainable* development is concerned with viable continuity into the long-term future. These differences, therefore, reach far deeper than UK environmental economists (Pearce *et al.* 1989, Helm 1991, Helm and Pearce 1991) allow for. They demonstrate that an environmental ethic is difficult to reconcile with economics of both the capitalist and communist kind. Insistence that economics and the environment are fundamentally linked and recognition that GNP is an unsuitable indicator of a country's economic growth, wealth and quality of life (Pearce *et al*. 1989) constitute significant first steps by (conventional) economists towards an acceptance of those incompatibilities.

Much bolder steps, however, are needed for a more radical environmental vision: environmental features of uncertainty, implication, invisibility, transience, temporal embeddedness and rhythmicity necessitate assumptions that allow us to relate to the globalized present, to historical and spatial extension, to the separation and reassemblage of time and space, to multiple time-spans, to speed, and to the gap between the time of information, machines and body movement across space. They require theories and metaphors that help us to extend simultaneously to the realm of the infinitely small and the unimaginably large: from nanoseconds to millenia and from the atomic and microscopic realm to the cosmos. If our contemporary environmental crises exhibit global features and are characterized by historically unprecedented characteristics such as ecologically networked interconnectedness, simultaneity and instantaneity, out-of-sync timeframes, multiple time lags, latency periods, and actions at a distance then, I suggest, these features must permeate our social science assumptions, our theories and our methods. They must become an integral part of our discipline in the same way as the key features of the Enlightenment had penetrated the work of the founding fathers of sociology and their successors. This, however, is difficult to achieve without significant changes to the discipline. Everyday life rather than theoretical tradition may be a potent source for inspiration.

In our everyday understanding we have no difficulty in coping simultaneously with those complexities and inconsistencies. We move with ease between long-term and daily concerns, the memory of friends we have not seen for many years and their altered appearance, the sense of urgency associated with a particular problem and its attendant future dangers. We interact with computers whose operating speeds are outside the capacity of human reaction and computation. We deal with historical matters that pre-date us by a multitude of generations and we function in a material reality of our own making which outlasts us by millenia. We relate as subjects not just to our fellow human beings but to our pet cat, special plants and even our car: we know their life-span of existence, their ailments and signs of decline. We receive in our homes events that take place on the other side of the globe and we are able to distinguish between an actual news report, a record of an earlier event, a future projection, a fictional dramatization and a computer animation. There is no difficulty about communicating through several media: by telephone, letter or on audiocassette. Nor is there a problem about distinguishing between the time of ageing, of the 'right' time for action, and the commodified time of market exchanges. All these times are equally available to us. We allow for the complexities to exist side by side, to interpenetrate and to inform each other. We cope with time lags and latency periods without necessarily requiring proof of the link between manifestation and cause, between friends from long ago and their present appearances, between the measles epidemic and our offspring's fever and rash. While our disciplinary tradition forces us to choose between both dualities and different temporalities, our experience combines the incompatible, relates different levels of reality and connects what is separated by time and space. We manage this because we do not rely exclusively on analyses based on sequential, linear causality, on 1:1 relations, and on traditional materialism. While we continue in our professional capacity to isolate, separate, dichotimize, fragment and abstract, in our personal lives we are able to relate, combine, bridge gaps and extend our concerns beyond personal and national boundaries. Moreover, we do not expect to control all the outcomes of our plans and actions. The complexity of everyday experience rather than perspective-driven tradition, therefore, is likely to provide the more appropriate guide to sociological engagement with present environmental crises.

As sociologists we need to overcome the clock-maker's reductionist view of nature and society and recognize our implication in the subject matter. We need to move ourselves from the outside to the midst of things, to become re-embedded. We need to be not just observers but participants who recognize that every action has effects, if not necessarily in the same time and place. Once we take account in our analyses of the complexity of everyday experience, unify the personal and professional realms and move from the illusionary position of objective observer to implicated participant, we will find it less difficult to extend our concern to the millenia by which the scientific impact on nature will outlast us.

SUMMARY

Running out of time, invisibility, and time lags spanning from nanoseconds to millenia pose unprecedented problems for theory and action. Moreover, sociopolitical engineering and risk calculation become highly problematic social practices when the projected disasters will strike before the traditional accumulation of data can provide clues to cures and when the potential outcome of cultural activity is characterized by globalization and temporal uncertainty. Traditional methods present barriers to effective action because our capacity to predict and control is dependent on processes governed by sequential, linear causality. This means that instantaneity and non-linear processes elude our conventional modes of domination: the simultaneous and potential present of global information and environmental effects render our efforts ineffective. The global present demands that prediction, action and control be based on different temporal principles. It necessitates response speeds outside the range of our physiological and mental capacity. It requires personal and collective responsibilities that span not just election cycles – or, at best, a single generation – but hundreds of generations. It calls for a bringing together of the time-scale of problems and the time-spans of concern. It links our personal/local times with global times and connects our individual concerns with collective responsibilities for the long-term future.

Explicit focus on time, therefore, gives us a new access to the contemporary environmental crisis. It illuminates the shadow side of environmental phenomena, aspects of processes which are normally ignored. It enforces a re-vision of the mutual impact of nature and culture and demonstrates the inappropriateness of traditional, perspective-based analyses. It suggests points of departure, particularly the need to let go of our exclusive reliance on the past as guide to knowledge for the future, and the necessity to change location and temporalities from external, static observer to involved, active agent. Once we shift position from exterior explicator to implicated participant, the environment no longer constitutes the 'other': it becomes an extension of physical, living, cultural selves, networked in never-ending relations. Importantly, this is achievable without losing sight of the workings of political interests and economic power. Finally, it involves social theorists in the transgression of disciplinary boundaries, the transcendence of dualisms and a coming to terms with uncertainty: the very approach necessary for a theory adequate to environmental matters. Environmentally engaged and time-based social theory converge.

ACKNOWLEDGEMENTS

I would like to thank Jan Adam, Dawn Clarke, Martin Read and Michael Redclift for their perceptive comments on an earlier draft of this chapter.

NOTES

1   Those differences and continuities are explored in detail in my *Time and Social Theory* (Chapters 3–7). Here I can only offer the briefest of summaries.
2   This time–energy link applies most specifically to traditional technologies but much less so to microchip information technology.

REFERENCES

Adam, B. (1990) *Time and Social Theory*, Cambridge: Polity; Philadelphia: Temple UP.

BBC Radio 4, 15 April 1992, 7 p.m.

Beck, U. (1992) 'From industrial to risk society: questions of survival, social structure and ecological enlightenment', *Theory, Culture and Society* 9: 97–123.

Becker, E. (1973) *The Denial of Death*, New York: Free Press/Macmillan.

Black Report (1984) Investigation of the possible increased incidence of cancer in West Cumbria (Report of the Independent Advisory Group, chairman Sir Douglas Black, DHSS) London: HMSO.

Brown, L.R. (1991) 'The New World Order', *The State of the World 1991*, London: Earthscan; pp. 3–21.

Churchill, R. (1991) 'International environmental law and the United Kingdom', *Journal of Law and Society, Special Issue: Law Polity and the Environment* 18: 155–73.

Cotgrove, S. (1982) *Catastrophe or Cornucopia. The Environment, Politics and the Future*, Chichester: John Wiley & Sons.

Dobson, A. (1990) *Green Political Thought*, London: Unwin Hyman.

Ekins, P. (1992) *A New World Order. Grassroots Movements for Global Change*, London: Routledge.

Helm, D. (ed.) (1991) *Economic Policy Towards the Environment*, Oxford: Blackwell.

—— and Pearce, D. (1991) 'Economic policy towards the environment, an overview', pp. 1–24 in D. Helm (ed.), *Economic Policy Towards the Environment*, Oxford: Blackwell.

Holliday, F.G.T. (1992) 'The dumping of radioactive waste in the deep ocean: scientific advice and ideological persuasion', pp. 51–65 in D.E. Cooper and J.A. Palmer, *The Environment in Question. Ethics and Global Issues*, London: Routledge.

Journal of Sustainable Forestry (1991) 1:1–123.

Macgill, S.M. (1987) *The Politics of Anxiety. Sellafield's Cancer-link Controversy*, London: Pion.

Maser, C. (1991) 'Adaptable landscapes are the key to sustainable forests', *Journal of Sustainable Forestry* 1: 47–59.

May, J. (1989) *The Greenpeace Book of the Nuclear Age. The Hidden History. The Human Cost*, London: Victor Gollancz.

Milton, K. (1991) 'Interpreting environmental policy: a social scientific approach', *Journal of Law and Society, Special Issue: Law Polity and the Environment* 18: 4–18.

Morin, E. (1974) 'Complexity', *International Social Science Journal* XXVI(4): 555–82.

Newby, H. (1991) 'One world two cultures: sociology and the environment', BSA Bulletin *Network*, vol. 50 (May): 1–8.

Palmer, J.A. (1992) 'Destruction of the rain forests: principles or practices?', pp. 81–94 in D.E. Cooper and J.A. Palmer, *The Environment in Question. Ethics and Global Issues*, London: Routledge.

Pearce, D., Markandya, A. and Barbier, E.B. (1989) *Blueprint for a Green Economy*, London: Earthscan.

Porritt, J. (1984) *Seeing Green. The Politics of Ecology Explained*, Oxford: Blackwell.

Postel, S. and Ryan, J.C. (1991) 'Reforming forestry', pp. 74-93 in L.R. Brown (ed.), *The State of the World 1991*, London: Earthscan.

Pugh, D. (1989) 'Getting into deep water', *Guardian*, 10 November, p. 29.

Renner, M. (1991) 'Assessing the military's war on the environment', pp. 132-52 in L.R. Brown (ed.) *The State of the World 1991*, London: Earthscan.

Romanyshyn, R.D. (1989) *Technology as Symptom and Dream*, London: Routledge.

Roszak, T. (1992) 'The voice of the earth', *Resurgence*, no. 150: 4-6.

Warren, L. (1991) 'Conservation – a secondary environmental consideration', *Journal of Law and Society, Special Issue: Law Polity and the Environment* 18: 64-81.

Yearley, S. (1991) *The Green Case. A Sociology of Environmental Issues, Arguments and Politics*. London: HarperCollins.

Young, J.E. (1991) 'Reducing waste, saving materials', pp. 39-56 in L.R. Brown (ed.), *The State of the World 1991*, London: Earthscan.

6

GENDER ANALYSIS AND ENVIRONMENTALISMS

Cecile Jackson

INTRODUCTION

The coupling of women and environment in development discourses, popular, academic and practical, has created an illusion of gender awareness. Yet women and gender are, of course, distinct, and this chapter aims to examine this illusion more closely. I focus on assumptions about women and environments but also raise the wider question of coercion in environmental management and regulation. A secondary theme of this chapter is to query the adequacy of the view that poverty is the cause of environmentally unfriendly behaviour. This leads to assumptions that poverty alleviation will result in more positive environmental management, and that therefore development and conservation are inherently compatible. A gender perspective, however, suggests that environmental behaviour is also formed by other social relations which can disrupt such an equation. It also suggests that environmental conservation is frequently predicated upon social inequality.

Development agencies from the World Bank and international organizations to bilateral agencies and non-governmental organizations (NGOs) have stressed in particular the role of women in synergistic and 'win–win' interventions which are seen simultaneously to advance women's gender interests and meet objectives of environmental conservation (World Bank 1992, Jackson 1993). Ecofeminist writers such as Vandana Shiva agree: 'women and nature are intimately related, and their domination and liberation similarly linked. The women's and ecology movements are therefore one, and are primarily counter-trends to a partriarchal maldevelopment' (Shiva 1989: 47). In addition, much of the women and environment literature[1] sees women's agency as confined to environmental friendliness and it is assumed that women relate to the environment in a positive manner except when forced by poverty to do otherwise. Thus interventions in support of the environment will benefit women, and development agencies are urged to involve women in conservation as fully as

113

possible. For the World Bank women have become the means by which environmental ends are achieved and part of this process has involved the manipulation of meaning in the vocabularies of gender and development analysis. The policy trade-offs and conflicts in the population–gender–environment nexus that contradict win–win assumptions are examined elsewhere (Jackson 1993). Here I focus upon the corroboration of synergism which is found in environmentalisms and suggest that this reflects conceptual problems, enthnocentrism, gender blindness, the naturalization of women and feminization of nature, and the absence of historical and material analysis.

Following a conventional recognition of at least two main strands of environmentalism, technocentric and ecocentric, I examine the gender content of sustainable development as a major thrust of techno-centric development in the South, before questioning the under-standing of gender in ecofeminism, and the degree to which it is reasonable to assume that indigenous peoples express a gender-equitable environmentalism. But first we we need to explain what gender analysis means in the context of the South.

Characterizing gender analysis

What we here call gender analysis[2] is a framework for understanding social change through the lens of gender identities. It has its roots in Western feminisms (Stamp 1989: 12–19), but is also influenced by other intellectual traditions, including development theories, and needs to be distinguished from feminism. The contribution of feminism to gender analysis is a major one – feminism has dismantled the idea that any particular class can represent the interests of humanity, documented the differential interests and power relations *within* domestic groups, shown how reductionist are notions of what con-stitute work and workers, and asserted that 'the personal is political'. Yet the divergences are significant.

Without engaging with diferent strands of feminism, one can recognize two ideas common to all: that women are oppressed at all levels of society and that conscious political action is needed to change this situation. It follows that there is a recognizable body of women's interests which are the objectives of this struggle. Feminist politics depend on the existence of male domination. Whilst the existence of women as a social category is fundamental to feminism, in gender analysis the notion of socially constructed and variable gender iden-tity is central. Gender is a culturally specific and socially conditioned identity of men and women – it is not fixed nor is it biologically deter-mined (Sayers 1982). The attention given by social science researchers working in the South to differences within the category 'women' has

114

perhaps follows from the especially sharp divergences of race, ethnicity and culture there, and it is no accident that in the West, feminist assumptions of the unitary interests of women and the primacy of gender subordination have been challenged by black feminists who point out that race completely transforms gender. In the feminist anthropology, or what we here call gender analysis, laid out by Moore 'there can be no analytical meaning in any universal conditions ascribed to . . . "woman" – for example, in the "universal subordination of women" and the "oppression of women"' (Moore 1988: 189).

The assumption of unitary interests of women is strongest in radical feminism and weakest in socialist feminism. Radical feminists are most willing to accept biologically based difference, as is ecofeminism which has strong linkages with radical feminism. Approaches to menstruation are a litmus test identifying differing forms of feminism, with radical feminists seeking not to demystify menstruation beliefs but to revalue them, for example, by the Baltimore Bleed-In. Thus radical feminists have suggested that menstruation is an 'heroic ritual shared by the community of women, connecting them with the rhythm of nature and with each other' (Delaney, Lupton and Toth 1976, quoted in Sayers 1982: 119). However, although feminisms have varying analyses of the bases of a universal 'sisterhood' its existence is a defining quality of feminism, but not of gender analysis.

It has been argued that there is a convergence of gender analysis and feminism (Bunch and Carrillo 1990) but the divergences, in development discourses, seem more striking. Unlike radical feminism gender analysis lacks the critique of male Western rationality, is not explicitly anti-intellectual, and has not generally questioned science, and unlike (much) Marxist feminism does not glorify indigenous, pre-colonial gender relations. The critique of the concept of patriarchy as a universal and generalized power of men over women is also a characteristic difference between feminisms and gender analysis, which prefers the concept of gender relations to that of patriarchy since the latter tells us little about how such control came into being, how it reproduces itself, what variations it displays and the gender struggles which shape it. Finally, gender analysis, unlike feminism, is less susceptible to the 'politics of the personal' slipping into essentialism and orientalist assumptions about indigenous peoples and nature.

Gender analysis emphasizes the importance of analysing women, as well as men, in relation to each other rather than in isolation, and of understanding gender relations at all levels of social organization. Since men and women characteristically occupy different positions in processes of production and reproduction they are affected by these processes differently. The importance of understanding gender divisions of labour is not only related to gauging the amount of work done by men and women, but also to recognizing that men and women do different work. However there are limits to what we can learn from the gender division of labour, and too often assumptions are made on the basis of this alone which fail to recognize that labour and responsibility are *not* the same. Gender divisions of rights and responsibilities, incomes, knowledges and decision-making are all constituents of a gender analysis. Gender ideologies are implicated in all these elements and form an important part of a gender analysis. In addition to these more 'structural' elements, a gender analysis suggests an understanding of the processes by which gender is created through the actions and choices, performances, struggles and strategies of individual women and men.

If gender analysis is a methodology with variable outcomes, then it is inappropriate to suggest a universal position on the degree to which women's gender interests and those of environmental conservation are compatible. However, gender analyses would appear to suggest a clash, more frequently than a complementarity, of environmental interests and women's gender interests. This chapter aims to demonstrate this opposition, without implying its inevitability, through a gender critique of different environmentalisms. I am not suggesting that there are no points of leverage which can address women's gender interests in development as well as environmental conservation. But I do aim to show that there are no grounds for believing that these are *necessarily* coterminous and that historical, social and ecological variations preclude such generalizations and necessitate site-specific gender analysis.

First to clarify two key terms. Definitions of environmentalism reveal two meanings – the first, which we shall not address, is the sense of environmental determinism. This is how the term is used in much anthropological literature (Ellen 1982). The second meaning centres upon 'the ideologies and practices which inform and flow from a concern with the environment' (Johnson 1981, cited in Pepper 1986: 13), and this is the meaning used here. Gender interests is a more problematic term. We avoid the term women's interests in order not to imply an essentialist view of women as an homogeneous category – gender interests will vary in time and space depending on how gender identities are socially and historically constructed and reconstructed. Some feminist theorists have also rejected the notion of interest because 'the language of interest, with its utilitarian connotations and connections to the "rational calculus", can never be redeemed to serve feminist purposes' (Jones and Jonasdottir 1985: 3), but here we sidestep the many difficult issues around understanding and revising the concept of interests and simply use the term gender interests to refer to 'those that women [or men, for that matter] may develop by virtue of their social positioning through gender attributes' (Molyneux 1985). We will use a gender lens to examine different strands of environmentalist thought and practice – technocentric, ecocentric and non-Western – to weigh up the validity of the claims for positive women and environment synergism.

SUSTAINABLE DEVELOPMENT

O'Riordan (1981: 1–19) distinguishes between two major strands in Western environmentalist thinking, the technocentric and the ecocentric, although in practice these often blend. Environmental regulation, reformism rather than radicalism, the scientific approach, belief in economic rationality and in the possibility of environmental management are characterisitc of technocentric environmentalism. Ecocentric environmentalism on the other hand is radical, utopian, romantic and characterized by its bioethical standpoint – that is, nature is valued not only for its usefulness to mankind.

Technocentric approaches to nature have been the dominant paradigm for some centuries and currently they involve instrumental and managerial resource-use planning (for example, environmental impact assessment) and conservation interventions. For Carolyn Merchant (1992: 2) it was the scientific revolution of the seventeenth century which asserted the domination by *man* of nature. Exploitative attitudes to nature, the powerful mythology of the scientific method and the deepening control and domination of women, developed simultaneously and Merchant contrasts this with the organic conceptions of nature and gender in earlier times. She suggests that the sixteenth century organic metaphor of a female earth, binding together humanity and nature, was replaced by the conception of nature and women (for example, witches) as threatening disorder and in need of control. However, it has been pointed out (Plumwood 1986: 126) that women were subordinated before the seventeenth century, as indeed was 'nature', and some have blamed the anthropocentrism of Christianity for the idea of a separate and dominant humanity apart from nature (White 1967: 1205). Whether this is so or whether, as some have argued, Genesis commits humanity to a stewardship, rather than dominion, over nature, the distinction between the transcendant and the immanent has been a consistent pair of opposed categories in Western European thought, in which male transcendalism was valued above female immanence.

The ideology of scientific and technological progress may not be responsible for initiating the exploitation of women and nature, but it appears to have contributed to these processes. The mythology of the scientific method as objective, neutral and value-free has veiled

117

the degree to which technocentric environmentalists and environmentalism expressed particular class and gender perspectives in what were perceived as problems, in the diagnosis of causes, and in the remedial actions taken. Technocentric environmentalism has no space for the politics of gender interests – environmental problems can be managed with technical means and economic incentives. It has no awareness of its own ideological stances nor of the impossibility of assumed gender neutrality in resource relations. The idea of science as objective and value free has masked the gender differentiation and biases in the recognition and prioritization of environmental problems, in access to political and bureaucratic power, in access to resources, environmental knowledges and information, in the valuation of diferent forms of work. If, historically, technocentric environmentalism has been gender-blind it will offer no explicit position on the question of positive synergism between women's gender interests and those of environmental conservation. In the absence of any gender awareness in technocentric discourses we can examine whether the *outcomes* of a gender-blind technocentric approach offer any support for the idea of synergism. If technocentric environmentalism has had little awareness of the gender-differentiated causes and consequences of environmental degradation what have been the effects on women of gender-blind conservation technologies in the South? Do they support the view that environmental protection and conservation is necessarily gender-equitable.

Many contemporary examples of the negative consequences of technocentric environmentalism for women are available (for example see Venkateswaran 1992, Agarwal 1986a, Fernandes and Menon 1987) to show that we cannot assume that women benefit especially (or even equally) from technocentric environmentalism. Typical development interventions with explicitly environmental objectives have included fuel-efficient stoves, solar cookers and biogas plants, agro-forestry, and soil and water conservation programmes. Social forestry and other tree planting schemes have been criticized for gender bias – for example, tree species chosen are the preferences of forest officers or men rather than women (Agarwal 1986a: 120), assumptions are made that women will care for the young trees (Hoskins 1983) or agroforestry crops (Leach 1991) without consideration of their opportunity costs, and gendered property rights are ignored (Bradley 1991) in afforestation projects.

Appropriate technologies have fared little better. Stamp (1989: 59–61) explains the slow and uneven acceptance of improved stoves in Kenya with reference to their unsuitability for cooking local food, and the problems deriving from loss of light and smoke and the immobility of the stoves. In Zimbabwe, the additional labour of women in tending fuel-efficient stoves, and the safety risks which precluded management by children were amongst a number of problems which reduced their potential for alleviating women's domestic drudgery (McGregor 1991: 470–6). Biogas plants have also been designed and implemented without attention to their acceptability to women or to their impact upon gender divisions of labour (Fernandes and Menon 1987: 139). Attempts to conserve and regenerate African rangelands have frequently involved the restriction of movement of pastoralists and herds. The group ranches created for Masai in Kenya has led to the privatization of pastures, a process which excluded women and consolidated male property rights (Joekes and Pointing 1991: 5). In soil conservation programmes women's unpaid labour has been a key ingredient (Thomas-Slayter 1992).

Beyond the particular forms of technical change pursued as strategies of green development lies what is articulated as 'sustainable development'. Although definitions are numerous there are some common elements –stewardship of natural resources and the rights of future generations, food sufficiency, consonance with 'tradional' cultures and livelihoods, environmental accounting, greater equity between North and South and within regions and popular participation (see for example, Conway and Barbier 1990, Davidson and Myers 1992). Sustainable development discourses certainly give more attention to women than earlier development approaches but this does not guarantee any more favourable outcomes for women, nor, as was pointed out above, does a recognition of women constitute a gender analysis.

The justification for special attention to women in sustainable development literatures is primarily based upon two views: 'Putting the poorest first also means prioritising the rights and needs of women' (Davidson and Myers 1992: 23), and that gender divisions of labour make women both responsible for environmental management and most vulnerable to the impact of environmental degradation. The first view, that poverty justifies a focus on women, is unsatisfactory because it collapses poverty and gender into a single phenomenon, yet we know that these are distinct axes of differentiation. Further, this view forgoes the opportunity to demonstrate the significance of social relations other than class to environmentally damaging behaviour and thereby challenge the simplistic explanation of poverty as a single cause of degradation (Blaikie 1985, Leach and Mearns n.d.). In practical terms this view rests the case for gender upon a demonstration of poverty, and implies that gender is irrelevant where poverty is not 'feminized'.

The justification which rests upon a stylized gender division of labour in which women are responsible for all reproductive labour, food production, domestic labour, child-rearing, community and natural resource management, and in which men monopolize cash cropping and market-oriented production, is seriously flawed for a number of reasons. There has been a trend towards the exaggeration of women's subsistence agricultural labour (Whitehead 1990, Tshibaka 1992, Dixon-Mueller 1985) and community management activities (Moser 1989), as well as a failure to understand the differences between divisions of labour and rights. This stereotype of rural women as excluded from commodity production is both false and dangerous in that sustainable development is poised to exploit women's unpaid reproductive labour further through the mistaken assumption that they are major stakeholders in conservation development. It is also dangerous to neglect the evidence, in many studies of commoditization, that engagement with markets has been progressive for women. The emerging picture of the Green Revolution experience for women of various classes shows that, for many, it produced an absolute rise in the work opportunities available to women, improved consumption and reduced domestic drudgery (White 1992, Lipton and Longhurst 1989, Sen 1985).[3]

Primary environmental care (PEC) has been suggested as a means of achieving sustainable development. PEC utilizes popular empowerment to secure basic rights and needs whilst caring for the environment (Davidson and Myers 1992: 27), gives a high profile to the role of women, and is promoted by the Overseas Development Administration (ODA), Oxfam and the International Institute for Environment and Development (IIED). It reflects the current enthusiasm for 'participation' and community based approaches to the environment. Whilst less technocratic than other elements of sustainable development practice, it suffers from a tendency to portray rural communities as homogeneous and harmonious and fails to understand the limitations of 'participation' from a gender perspective. Participatory approaches to development assume that communication is unproblematic and ungendered, beyond the need to make sure that women and men are represented in the decision-making or consultative bodies involved. There is no recognition of the degree to which views expressed by participants reflect dominant/dominated models and knowledges, 'false consciousness', or mutedness. What is said cannot be taken at face value. Women may devalue themselves in line with dominant gender ideologies, they may be unable to articulate alternative views, or unwilling to do so where these may generate gender conflict. Empowerment does not necessarily follow from participation – particularly for women – because participation of women does not necessarily express their gender interests. In addition, PEC implies an empowerment of a community which is conceptualized as internally undifferentiated and sharing joint interests and fails to see that what may empower poor men may not empower poor women. Women are portrayed in PEC as altruistic actors – a portrayal that allows community work to come to mean voluntary work by women.

Sustainable agriculture has developed a considerable momentum within the range of farming systems research (FSR) perspectives which emerged in the 1970s and became widely institutionalized, both internationally through the Consultative Group on International Agricultural Research (CGIR) system and within the national agricultural research systems of developing countries. FSR is based upon an interdisciplinary, whole-farm systems approach to technology development for, and with, resource-poor farmers. These characteristics raised hopes that FSR would be an appropriate vehicle for the incorporation of gender analysis into sustainable agricultural development (Poats *et al.* 1988). However this has not materialized, despite the growing volume of FSR work with women farmers, because social scientists have remained marginal within FSR teams and because FSR in practice has continued to focus on crops and enterprises prioritized by male farmers and researchers. And 'In practice the FSR/E approach is used primarily to reduce the degree of error in finding the appropriate technical fix for market-based problems of agricultural production' (Evans 1988: 39). Thus the reproductive labour of women, unpaid, may be value-adding but is treated as costless in the evaluation of production bottlenecks or of alternative technical solutions. Furthermore, although rural livelihoods are seldom based upon agriculture alone, FSR has not satisfactorily incorporated this broader perspective – for example, the constraints of domestic labour or the opportunities of remittances from wage workers.

Despite the volume of research on gender roles in rural livelihoods FSR researchers have failed to follow through the implications of this research: most importantly, the critique of the nature of the household. FSR can integrate women as worker-members of farm households but is not good at dealing with complex differentials in costs and benefits to men and women within households where there is split respon- sibility and decision-making powers.

The debate about the nature of the household which has developed from feminist scholarship has significant implications for environ- mentalist thinking and practice as well as for development more broadly (Guyer 1981). The household has been conceived of as a unitary body with a range of functions – production, consumption, residence, reproduction, and so on – in concepts and models as well as in descriptive empirical work, development policy and practice. In recent years this has been challenged by alternative views of households as having different forms and functions according to class and other social divisions (Netting *et al.* 1984), as well as stages in the development cycle of the household (Goody 1971) that is, in the processes of household formation, expansion and dissolution. But the critique of the unitary nature of households has been made most strongly from the perspective of the conflicting interests of men and women within households.

From our discussion of gender divisions of labour, rights and responsibilities we can see some of the ways in which men and women may have distinct and different interests although they are members of the same household. The question of how differing self-interests are reflected in decision-making is important and raises directly the issue of power in decision-making. Economic models have sought to overcome the problem of how individual utilities of household members can be aggregated to a joint utility (theoretically inadmissible) by assuming the existence of a benevolent dictator who makes decisions in the interests of the household as a whole (Folbre 1986). It has not been difficult for gender analysts to show that household heads (predominantly male) do not control all decision-making, that they do not necessarily arbitrate fairly in the collective interest, and that gender conflict of interests and outcomes are present to varying degrees in most if not all societies (Folbre 1986, Bruce 1989). Bargaining models and approaches which integrate both co-operation and conflict within households (Sen 1987) indicate the indeterminate nature of decision-making processes, the central role of power, the impact of wider societal level factors upon decision-making and the importance of subjective perceptions (of self-interest, of self-worth and of labour value, for example) in patterning outcomes. From a gender perspective this sets gender relations centre-stage in any attempt to explain household-level decision-making with regard to environmental management.

It also implies that individuals within households will have differing objectives and livelihood strategies – some will be shared but others will conflict. This is a central issue which sustainable agriculture initiatives fail to consider. For example, in the promotion of low-external-input agriculture it is recognized tht households are composed of individuals with differing decision-making spheres, but it is then asserted that household members share common objectives, and thenceforth 'farmers' and 'households' are used synonymously (Reijntes *et al.* 1992: 29–31).

It also remains true that sustainable agriculture pays only passing attention to social issues and is dominated by technical approaches. The low-input strategies for sustainable development may reduce the need for external (that is, purchased) inputs to agriculture, but in so doing increase the use of 'free' inputs like women's labour. In the context of southern Africa the adoption of high-yield varieties of maize would appear to be related to their lower labour demand in cultivation (Low 1986), and a promotion of traditional varieties may involve women in greater unpaid farm work and lower productivity. Many of the cunning techniques to sustain soil fertility or conserve water are very labour demanding and it is possible that low-input agriculture leads to deepening exploitation of women's labour at the household level which parallels the impact of PEC at the community level.

To summarize this section, technocentric environmentalism is largely gender-blind, either because it fails to recognize gender differentials at all, or, where women are recognized as a distinct category, because gender stereotypes prevail and the household continues to be treated as a unit. This is not to say that women are passive and inevitable victims of technocentric environmentalism or that they do not benefit at all from it, either directly or because of unintended consequences. However, we can say that conservation technologies are not inherently favourable to women, let alone synergistic with their gender interests.

The intellectual currents discussed in the next section which feminize nature and naturalize women do influence technocentric environmentalism – the boundaries drawn here between technocentric and ecocentric are not impermeable – and furthermore the vacuum of gender awareness in technocentric environmentalism has allowed assertions of synergy to pass unchallenged, since gender conflicts have been invisible and unrecorded. But if gender-blind technocentric environmentalism somewhat predictably fails to demonstrate a coincidence with women's gender interests it may be reasonable to expect that ecofeminism will.

ECOFEMINISM AND INDIGENOUS ENVIRONMENTALISMS

Ecocentric environmentalism can be traced to European romanticism of the seventeenth and eighteenth centuries and to nineteenth-century American transcendentalism. In this reaction against rationalism and science, alternative knowledges based on feelings, emotions, instincts and morals were revalued. Although perhaps still a secondary stream in environmentalism, ecocentrism is of growing significance – a theoretical literature is developing rapidly (Merchant 1992, Eckersley 1992, Scarce 1990), as are a diverse array of ecocentric activist movements, such as Greenpeace, which share a bioethical standpoint. Ecocentrics have been quick to claim that they have inherited the mantle of emancipatory thought (Eckersley 1992) from socialism, and that anthropocentrism/ecocentrism replaces left/right as the most relevant political cleavage.

Although ecofeminism is still relatively recent one can identify some common themes in ecofeminist literature.[4] Ecofeminists see women as closer to nature than men, they oppose the domination of nature by humanity and insist that nature has no hierarchies which are seen to be derived from hierarchical human societies and imposed upon nature. A bioethical view, that forms of nature are not of differential value, is sustained in much ecofeminist writing. Although we examine ecofeminism as an example of ecocentric thinking it would be mistaken to assume that there are no differences within ecocentric groups. Ecocentric selflessness has been a divisive element in ecofeminism – resisted by some (Biehl 1988) to the point of renouncing ecofeminism, and embraced by others such as spiritual ecofeminists.

Ecocentrics aim at reducing human numbers because population growth is seen to 'magnify environmental degradation and therefore impair the overall quality of human life' as well as to have negative impacts on 'the nonhuman community' (Eckersley 1992: 130), and they support population control in the Third World and restrictions on migration from the South to the North. Movements such as Earth First! are deeply Malthusian – the journal of Earth First! declared that 'if the AIDS epidemic didn't exist, radical environmentalists would have to invent one' (quoted in Merchant 1992: 175). It is difficult to tell what ecocentric theorists like Eckersley mean when they promote 'humane birth control' in the Third World, but what is evident is the absence of discussion of population programmes which have infringed women's human and reproductive rights – such as the sterilizations in Bangladesh (Hartmann and Standing 1989) – or what bioethics prescribes with regard to abortion.

We will now concentrate on an examination of ecofeminism as an example of ecocentrism in our questioning of how far we can see women's gender interests and environmental interests as compatible. This section is structured around the ecofeminist acceptance of the woman and nature connection, the inherent essentialism in ecofeminist views, and the ahistorical nature and absence of material context in ecofeminism.

Two dichotomies are basic to the cofeminist position: nature and culture are opposed categories, as are women and men. The formulation that men are to culture as women are to nature is accepted. However, both these dichotomies have been shown to be untenable by MacCormack (1980: 17). Culture is grounded in the human brain which is part of nature; *both* men and women are physically part of nature and also have mentality and therefore are part of culture. Another problem is that the nature/culture dichotomy cannot be

124

ethnocentrically universalized from Western intellectual history. Naturalizing women, and gender relations, has been seen by Western feminists since de Beauvoir (1988) as a means of justifying gender inequality and therefore to be resisted. Ecofeminists, however, rather than resisting the idea that women are linked with nature, celebrate and revalue this linkage. They see the women–nature connection as a vantage point in the struggle for new ways of relating to nature that are not characterized by domination and control. They revalue the spiritual, the intuitive and the instinctive as alternative, and superior, forms of knowledge to science and rationalism (King 1989).

Ecofeminist acceptance of the women–nature link rests upon the roles of women in biological reproduction and the psychological and social conditioning (Ortner 1974) which are seen to follow upon this – women's bodies are creative and productive like nature; through child-rearing they come to develop caring, nurturing and altruistic behaviour which is extended to nature; they are not separated from their nurturing mothers nor therefore individualized and socialized into independence and autonomy as men are; and they lack the competitive and controlling impulses characteristic of man's relationship to nature. The biological determinism of this view has been discussed elsewhere (Sayers 1982, Brown and Jordanova 1982), but it is an unsatisfactory position because it cannot account for variation in reproductive and environmental relationships, or for how these change over time.

Ecocentric environmentalism suffers from essentialism in relation to both women and environments. Women are conceived of as a unitary category with universal characteristics which transcend the time, place and circumstances of their lives. Ecofeminist discourses are innocent of gender analysis in which masculinity and feminity are relational, socially constructed, culturally specific and negotiated categories. Ecofeminists, like radical feminists, seek to recognize and revalue the feminine. Thus, caring and nurturing are seen as universal feminine characteristics and as a model in remaking the relationship of humanity with nature. However, this presents a number of problems. Given that the feminine has formed in relation to the masculine, how is it possible to discover a feminine 'essence'? Ecofeminists such as Val Plumwood (1988) recognize that remaking humanity in a feminine form which cannot be known poses a serious problem to the ecofeminist project. Vandana Shiva attempts to overcome the problem by distinguishing the 'feminine principle' from actually existing women, but then constantly lapses into an elision of women and the feminine (Shiva 1989: 109). The empirical evidence for a view that women are universally closer to nature seems to be limited (MacCormack 1980). Women are sometimes perceived as closer to nature and sometimes as more associated with the domestic and the settled – that is, with culture (for example, Gillison 1980). One of the problems in researching this question is the validity of the basic nature/culture opposition for where, as we see in some of the examples below, nature is intimately bound up with belief and ritual the dichotomy breaks down.

The term 'nature' like 'women' also needs to be interrogated, for ecofeminists also essentialize nature and do not see nature and environment as culturally relative but as biological facts. Anthropologists (for example, Douglas 1973) and Marxists (Schmidt 1971) have long recognized that nature is expressed and known through symbols which are culturally specific. The meaning of nature is dependent on historically and culturally specific understandings which reflect gender differences as well as other social divisions. Nature is conceived of as female in Western culture (Mother Earth and earth mothers) and ecocentric environmentalists continue to develop this imagery – the name Gaia is taken from an ancient Greek earth goddess and environmentalists continue to use this female imagery (Murphy 1988). One of the most frequently cited passages in ecocentric discourse is the quote from the Red Indian, Chief Smohalla: 'You ask me to plough the ground: shall I take a knife and tear my mother's bosom? You ask me to cut grass and make hay and sell it to be rich like white men; but how dare I cut off my mother's hair' (Eastlea 1981: 43). The feminization of nature says more about the cultures and texts in which this dialogue appears than about any inherent and universal character of nature as female or any universal preindustrial equality of women.

The making of ecomyths which essentialize and romanticize both nature and non-Western peoples, especially women, is typical of ecocentrism. The construction of the East as spiritual and ecologically aware is a form of orientalism whereby agency and rationality are then, dichotomously, seen as the preserve of the West, and Guha complains that 'varying images of the East are the raw material for political and cultural battles being played out in the West; they tell us far more about the Western commentator and his desires than about the "East" ' (Guha 1989b: 77). One famous speech, in 1854, by Chief Seattle of the Suquamish tribe has been shown to be actually a third- or fourth-hand version and 'many of the words which resonate with modern ecological consciousness are not the original words, but contain phrases and flourishes designed to appeal to ecological idealism and the Christian religion' (Merchant 1992: 122).

Women have become part of this construction of the Other through the women–nature iconography of Western society. The Chipko movement in India, which is widely given as an example of spontaneous mobilization by women in defence of the environment, emphasizes the holistic understandings of women, the feminist character of the movement, and the altruistic motivation of the women. The Chipko slogan, 'What do the forests bear? Soil, water and pure air', is said to show the women's essential ecological understanding of hydrological cycles, and the use of the forests by women is referred to as traditional and ecological. Yet what is underemphasized is an analysis of the material and historical conditions which led Chipko women into environmentally protective behaviour (Guha 1989a), and a recognition that women typically over-exploit forest resources as much as men (Kelkar and Nathan 1991).

To summarize – the false idea of the positive synergism of women's gender interests and environmental interests seems strongly related to an essentialist denial of the social, and to a historical construction of gender and nature. We briefly examine these next.

History appears in ecofeminist thought in a largely linear 'before and after' (scientific revolution, colonialism) manner. Ecofeminist readings of history for different regions have some similarities. Both Merchant, for Europe, and Shiva, for India, conceive of harmonious complementarity in both gender and environmental relations: before the scientific revolution in the case of Europe and before colonialism in the case of India. For example, Shiva, in describing colonialism, writes that 'Maldevelopment is the violation of the integrity of organic, interconnected and interdependent systems, that sets in motion a process of exploitation, inequality, injustice and violence' (Shiva 1989: 5–6)

The feminine principle, which united people with nature in pre-colonial India is described by Shiva as 'the ancient Indian world-view in which nature is Prakriti, a living and creative process, the feminine principle from which all life arises' (Shiva 1989: xviii). However, it has been pointed out that Sanskrit texts from which such a world-view is drawn represent the views of rich, high-caste men (Rao 1991: 19). Further, Bina Agarwal has remarked that Shiva conflates the Indian with the Hindu (Agarwal 1991: 9) in her assertion of the feminine principle and thereby glosses over the plurality of ideologies and interests in pre-colonial India.

Shiva's representation of pre-colonial harmony and equality is highly questionable for both gender relations (for example, suttee in India or domestic slavery in Africa) and environmental relations. '"[M]ainstream" Indian civilisation was set up by subjugating the forest dwellers and clearing the forests for settled cultivation' (Kelkar and Nathan 1991: 112). Indian civilization has also been repressive to women:

> The absence of any seclusion of women in the tribal situation, the free mixing of adolescents of both sexes, the choice of women

and men with regard to their marriage partners, the ease of divorce, the practice of widow remarriage – all come under severe attack in the period of formation of caste society.

(Kelkar and Nathan 1991: 113–14)

Below we examine the degree to which indigenous environmentalisms are based upon ecocentric views and the absence of women's subordination.

Ecofeminist literature relies heavily on the concept of patriarchy, which is conceived of in a monolithic, ahistorical and reductionist manner. Within gender and development discourse there has been much critical debate about the usefulness of the concept of patriarchy which fails to distinguish the variations in mechanisms and structures of gender inequality, their changing character and their reproduction (Whitehead 1979). Ecofeminism reflects an awareness of the problem of how humanity relates to the non-human, and it reflects the struggles within feminism generally of how to construct the human in other than masculine characteristics. But an ahistorical essentialism cannot be the basis for such a project. Why are materialist perspectives so weak in ecofeminist discourse and what are the material circumstances which pattern gender roles in relation to environmental management?

In addition to essentialism the ecocentric insistence on transcending left/right politics, the identification of environmental degradation in Eastern Europe with socialism, and the feminist critique of Marxism are perhaps elements explaining the absence of materialist perspectives in ecofeminism. Left environmentalists can hardly be called Marxist given the degree of revision of Marxist thought, and the term 'radical' has been captured by ecocentrics. On the whole, Marxists have seen the development of the forces of production at the expense of nature as progressive and environmentalism as an elitist preoccupation – attempts to reconstruct Marx as an environmentalist are strained and selective. However, there are important insights in Marxist perspectives which are absent from mainstream and ecocentric environmentalism – for example, the unity of nature and culture; the social construction of nature; the manner in which material conditions and history inform environmental ideas and perceptions; the internal differentiation of society which leads groups to have different environmental relations. These are a necessary antidote to the fundamentalism of ecocentric environmentalism and the apolitical stance of technocentric environmentalism. What is missing though, from a gender perspective, is considerable – the emphasis upon class alone is inadequate, and intraclass divisions and conflicts considerable – the the method without the orthodoxy remains necessary.

A key problem with ecofeminist approaches is that they fail to recognize either the diversity of lived environmental relations which different women experience, or the power structures in societies which mediate environmental relations and the ebb and flow of competing environmental ideologies. Accounts of how class relations impinge upon resource access reveal considerable variations – for example, the poor may make particularly intensive use of commons, and levels of inequality may be reduced by access to common property resources in India (Agarwal 1991: 13). In other regions, though, commons may be captured by the rich and inequality deepened – for example, grazing lands in Botswana (Cliffe and Moorsom 1979).

Gender differentiation means that men and women of the same household relate to resources in different ways and these variations are inserted into class relations. But the outcomes are not predictable – poor women may be more or less environment-friendly in their behaviour than poor men or rich men/women, depending on their rights, responsibilities, knowledges and bargaining positions within their households and communities. Thomas-Slayter (1992) describes class-gender relations in Kenya which lead poor women to effective group soil-conservation activities but prevent them taking action over sand scooping which is seriously damaging water courses and availability. Attempts to introduce alley crop farming in Nigeria encountered low uptake amongst women farmers in the south-west because the *Leucena* planting time clashed with the peak oil palm processing period for women, and furthermore weeding and fodder cutting placed heavy demands on women's time. In the south-east of Nigeria alley crop farming was resisted by women – weeding was poorly done or crops such as melon were planted by women which smothered the seedlings – on the grounds that they were uninterested in the fate of 'men's' trees (Leach 1991).

The problem of unitary conceptions of households underlies the failure to perceive some contradictions generated by gender relations for environmental conservation. Livelihood strategies for men and women within a household vary and reflect gender relations – women may seek an autonomy which can imply harmful resource use. For example, beer brewing in much of southern Africa is both a means by which poor women recruit farm labour and an activity which for many women generates independent cash incomes; yet beer brewing requires large amounts of fuel and contributes significantly to deforestation. Beer brewing involves very long cooking periods and hence frequently leads to the cutting down of live wood since wood of large diameter is required, unlike fuel for food cooking which is generally collected as smaller dead branches.

129

In addition to class divisions women are disaggregated by age and life-cycle in their environmental relations – although wood and water are generally said to be collected by 'women' we find on closer inspection that not all women bear this burden equally and many older women manage to delegate this responsibility to sons' wives. Similarly, developmental cycle variations in household size and composition generate significant variations in the reproductive labour of women.

Why is it that ecofeminism has become so internationally influential? An example of this influence can be seen in the high profile given to ecofeminism at UNCED in 1992, and its assumptions which have passed so readily into the women and environment literature. A number of reasons spring to mind – the effectiveness of feminist critiques of Marxism and technocentrism, the crisis in the left, the ascendancy of ecocentrism especially in America, the influence of radical feminists in the European green movements.

But within development discourse this is possibly also related to the current strength of agrarian populism (Kitching 1982) in NGOs and other development agencies. In Britain this may reflect economic pressures in recent decades which have led the disaffected urban middle classes to deepen their preoccupation with rural life (Newby 1979), the glorification of an organic past, a concern with 'community', and non-material values. Both agrarian populism and ecocentrism are influenced by romanticism, and key figures (for example, Schumacher and Gandhi) have a central significance to both. Agrarian populists see peasants as undifferentiated, virtuous (Bernstein 1990) and co-operative; they demand participation and bottom-up approaches which reject hierarchy, the use of appropriate technology and self-sufficiency, and the recognition of indigenous technical knowledge. There appears to be a strong affinity here with ecofeminist emphases upon women as an homogeneous (and virtuous) category of altruistic and community oriented people, on webs rather than hierarchies, on utopian self-sufficiency, and on women as 'the intellectual gene pools of ecological categories of thought and action' (Shiva 1989: 46). Finally, women and nature may have proved an attractive and acceptable linkage for Western women environmentalists, for whom ecocentric values have seemed to offer an opportunity to reconcile the project of feminism with that of environmentalism.

The question of how far, and why, ecofeminism has gained adherents in Third World countries requires further study, and is problematic because of the difficulty of bounding cultures when discourses are now so profoundly globlized (can Vandana Shiva be taken to represent the South?); because of the manner in which Third World environmental movements are represented in development discourses dominated by Western culture (for example, Chipko, discussed on pp. 126–7); and because of the financial rewards for southern NGOs of conformity with the expectations of bilateral and multilateral development agencies.

Next we consider whether there are grounds for the synergism argument in what is known of Southern environmentalisms. The assumption that indigenous environmentalisms exist which are gender-equitable or which recognize a special connectedness of women and nature has to be interrogated, at least partly because of the growing emphasis in development practice on grass-roots initiatives for sustainable development.

Attention to non-Western environmentalisms has been slight – with the notable exception of the analysis of Indian environmentalisms by Ramachandra Guha (1989b). Taken as a whole, Indian environmentalism is markedly different to that of the North because it reflects competition over productive resources rather than leisure and quality of life issues; because India is not a post-industrial society; because environmental action is essentially an aspect of peasant movements in India; and because of the severity of the livelihood impact of resource degradation in India. Guha distinguishes three strands of thought in India – the crusading Gandhian movement, the Appropriate Technology movement and the ecological Marxists – all of which in some ways correspond to forms of environmentalism in the West. There are interesting similarities in Indian environmentalism to currents in Western environmental thought – the Gandhian element reveals a heavy moralism, a rejection of materialism, a call to return to precolonial harmony and a religious reverence for nature (Gadgil and Guha 1992a) which has parallels with ecocentrism. Ecological Marxists are, on the other hand, hostile to tradition, positive in their attitudes to science and industrialization and emphasize inequality in their analyses of the causes and consequences of environmental degradation. A distinctive form of feminist environmental analysis (Agarwal 1991) may well emerge from ecological Marxism but it is difficult to identify 'indigenous' models when globalization blurs all boundaries – Guha 1989a) identifies Marxist influences on movements like Chipko, and Western ecocentrics acknowledge Gandhian influences upon their thought.

How do folk models and cosmologies treat gender and are there conflicting interests here too? To begin to answer this question we need studies of environmental knowledge systems which go beyond folk classifications and reveal an explicit ideological stance of environmental protection – it is not clear how far such environmentalism itself is a Northern idea. What we can do is briefly examine concepts of nature and studies of spontaneous peasant mobilization

131

in defence of the environment for clues to the question of the complementarity of conflict in women's gender and environmental interests.

Nature is perceived in cultural terms, and understanding folk categories depends upon an understanding of particular cultures. Anthropological studies of folk classifications have limited usefulness for our purposes – what we are interested in is 'a more discursive indigenous knowledge of environmental phenomena and the technical theory and practice associated with them' (Ellen 1982: 210). We would like to know whether men and women of particular societies have different such knowledges and roles, whether these differences are recognized and whether those of women are more environmentally protective. The question of outcomes – that is, of whether women behave in a more environmentally friendly way – is of course different. Pursuing these questions is particularly difficult because gender has not been a focus in many studies, and there is the further problem of mutedness (Ardener 1975). What is known about women's environmental perceptions is largely collected from male informants, and even where this is not the case the domination of male worldviews and the absence of, or the politics of expressing, a female vocabulary to articulate dominated environmental models is a major hurdle in interpreting what women themselves say.

Studies of indigenous knowledge systems, such as van Leynseele's (1979) work in Zaïre, show that the cognitive system is closely related to the system of exploitation and regulation of the environment. In his study, overfishing was prevented by limiting habitable space (and thus overpopulation relative to the fish resources), by technical means (for example, fish basket gauge), and by private property rights to pools and fisheries. He concludes that

> the interventions made within the niche with a view to intensive production presupposes a genuine understanding of the [ecological] processes and a consciousness, at least implicit, of the consequences of overexploitation. The degree of understanding of the environment may be established by comparing the ecological system and the classification established by the specific terminology of the vernacular language . . . [which] reveals a recognition of the functional relations operative between elements of the environment as they emerge from the ecological processes.
>
> (van Leynseele 1979: 181)

The rights of an agnatic core of kin and the exclusion of others are central to this system, and the ecological understandings and management decisions appear to be held and made by dominant males. This is as one might expect; where survival depends on preventing resource over-exploitation, environmentalism is of central political significance and reflects other pervasive power relations – in this case, male dominated. Women here are not inherently destructive of, or indifferent to, environmental sustainability, but many may experience social relations which limit their power to form environmental knowledges, or the right to express them, and operate to exclude women from direct property relations. We also see from this example that it is a technocentric (that is, an instrumental) relation with nature which is revealed rather than a bioethical view: management in the interests of human survival was behind the system. Yet ecocentrics often claim, with the very selective and often inaccurate use of examples of non-Western societies, that the organic harmony of humanity with nature in such societies reveals a bioethical standpoint. Mary Douglas observed some time ago that such a 'universe is man-centred in the sense that it must be interpreted with reference to humans' (Douglas 1966: 85), and she emphasized the limitations of studies of cosmologies in isolation from practical concerns

> The live issue is how to organise oneself and other people in relation to them; how to control turbulent youth, how to soothe disgruntled neighbours, how to gain one's rights, how to prevent usurpation of authority or how to justify it. To serve these practical social ends all kinds of beliefs in the omniscience and omnipotence of the environment are called into play.
>
> (Douglas 1996: 91)

At the very least we need to interrogate the meaning of, say, the protection of certain species by indigenous peoples before assuming they are non-speciesist.[5] Religious taboos controlling resource use certainly do protect elements of environments, but a materialist analysis of what is protected and why, as well as who the major beneficiaries are, must precede any generalizations about the absence of speciesism.

One example of a representation of indigenous environmentalism taken out of context is the suggestion that menstruation taboos, in Orissa, India, are an expression of a unified conception of nature and culture in which '[t]he actions of humans must harmonise with the movements of the sun, of the clouds, with their convergence or separation from the earth. Women and men recapitulate in a monthly rhythm the earth's yearly rhythm' (Apffel Marglin 1992: 30). These taboos require women not to touch anyone during the first days of menstruation, not to cook or wash or bathe or have intercourse, to sleep on a grass mat and eat a very restricted diet. A man's life will be shortened by the touch of a menstruating woman. Marglin's interpretation is achieved by an illegitimate extrapolation of what villagers say about Raja Parba, the festival of the menses of the earth – in which women are left without men in the villages to amuse themselves – to actual menstruation behaviour. Thus she emphasizes the release from work for menstruating women. This is misleading – when a menstruating woman does not cook, another woman, not a man, cooks. The idea that menstruation taboos have a relation to male domination is dismissed, thus 'the understanding of menstrual taboos as signifying a male domination of women's sexuality and a way of keeping women out of the productive labour force amounts to inventing them as commoditised persons' (Apffel Marglin 1992: 26). She further asserts that freedom of choice is a 'commoditised logic' and that menstrual taboos serve to 'ensure continuity by articulating with the movements of other persons and of the seasons' (Apffel Marglin 1992: 31). We are given no analysis of class or caste or the position of women relative to men in general, nor any discussion of pollution and gender relations (Douglas 1966). Submission to taboos emanating from a religion formed by elite males is decontextualized and presented as a superior organic unity of humanity with nature.

It is necessary to see beyond the invention of environmentalist traditions of peasant societies by the recasting of ritual, religion and belief as indigenous ecocentric environmentalisms.

An example of the determination to represent even dehumanizing elements of indigenous cultures in an ecologically positive light can be seen in the reinterpretation of human sacrifice amongst tribal groups in India. Mahapatra (1992) asserts that the Kondhs, a tribal group of southern Orissa, live in harmony with nature. The evidence for this is the worship in the past of an earth goddess by human sacrifice, *meriah*. Mahapatra represents the *meriah* in a particular manner – he claims that the human sacrifice enabled Kondhs to belong to a place because their 'grandfather's bones are beneath the soil' (Mahapatra 1992: 65), yet the *meriah* was always a stranger, purchased at considerable cost from low-caste artisans and weavers (Leigh Stutchbury 1982). He justifies the sacrifice by stating that the *meriah* was a 'voluntary victim' and was drugged before being killed (quoting Frazer on this!), and he dehumanizes the *meriah* by calling s/he 'it' (Mahapatra 1992: 65–6), or in other places 'the object' (ibid.: 68). Historians have given other accounts of *meriah*. 'The method of immolation varied among the tribes, though in all cases the men cut the flesh from the living victim, who was usually drugged or intoxicated, or made defenceless by having both arms and legs broken' (Leigh Stutchbury 1982: 45).

Leigh Stutchbury compares the *meriah*-practising Kondhs with those who did not and finds that the latter group were characterized

134

by purity concerns, menstrual taboos, dowry, the absence of divorce and widow remarriage, hypergamy and very high rates of female infanticide.[6] The *meriah* Kondhs had few if any menstrual taboos, relatively high status of women, divorce and remarriage of women were common, and exogamous bride-price marriages the rule. She suggests that *meriah* was made to a female deity presiding over chaos in tribes where marriage customs failed to resolve the threat posed by women as the 'enemies within' exogamous societies (Douglas 1966), whereas for the infanticide Kondhs status hypergamy[7], similar to that of the Rajputs today (see Billig 1991), controlled both women and relations with outside groups. This example shows the problems of a literal reading of articulated beliefs. Earth goddess worship here expresses the anomalous position of women rather than 'a lifestyle . . . tied up with the land in a system of mutual reinforcement, . . . attuned to the sonic and sensual rhythms of the earth' (Burman 1992: 3). It also shows the degree to which issues of equity and human rights can be willingly overlooked in the claims made for indigenous environmentalisms.

An example of how indigenous environmental models may express gender conflict in a way that portrays women as antagonistic to nature can be seen in the study of the Dogon at the Mali/Burkino Faso border where van Beek and Banga (1992: 57–75) describe the punishment by men of women firewood collectors gathering valued fruit-tree wood. Women collect firewood, they brew beer and they fire pots – all these use substantial quantities of scarce wood. Yet the punishment is not just an expression of the conflicting individual self-interest of the women and men who wish to consume the fruit of the trees, it is also an expression of a kind of male environmentalism. Dogon attitudes to the bush embody respect for the powers of the bush, upon which the village depends – for the bush represents life and culture means entropy. However, although this leads to a wish to conserve the bush, 'working harder is a perfectly feasible solution to all environmental pressures' (van Beek and Banga 1992: 72) and there is no acknowledgement of the role of people in actively regenerating the bush. The punishment of the women involves the bush (actually men in masks) come into the village to seek out fruitwood in firewood piles and to chastise the women for their lack of respect by imposing fines on women, which are however paid by their husbands. 'Through these rituals men have appropriateed both the life-giving and the life-threatening aspects of the bush' (van Beek and Banga 1992: 73). What is interesting here is that it is men who are associated with nature; women are seen as subverting respect for nature; respect for nature is inextricably bound up with respect for men by women, and finally the payment of the fines by the men indicates perhaps the ambiguity of real gender power-relations and the acceptance by men of joint responsibiity for deforestation.

If environmental relations are patterned by gender, how far do they also refract other relations of social inequality? Research (van den Breemer 1992) in Cameroon shows women again to be associated with settlement and culture and men as the mediators with the wild. But this is incidental, the main finding in this work is that 'ecologically destructive processes have their roots in the aspiration towards emancipation' (van den Breemer 1992: 106). The religious prohibitions against ecologically damaging crops and animals (rice and goats) become overthrown as a result of the contradiction between these ideas about the socio-ecological order and the system of ideas which legitimizes hereditary leadership and social inequality. Matrilineal hereditary leaders required labour as power-bases and they built up followers of immigrating foreigners and (ex)slaves who then, in the 1920s, converted to Islam and Christianity (and therefore were absolved from traditional religious observances) and took up opportunities for cocoa and coffee cultivation, which was individualizing and emancipatory. This study does not have explicit gender analysis but it affirms the point that environmental conservation is often predicated upon social inequality, and that the emancipation of women (or other dominated groups) may create inevitable breakdowns of eco-order.

The degree to which gender relations are fundamental to understanding patterns of resource degradation (for example, deforestation) in the South is illustrated by a study of the Susu of Sierra Leone. The concept of wealth-in-people has been used to understand social relations in agro-ecosystems where population densities are low and labour constraints are a major feature of rural livelihoods. A number of writers (notably Meillasoux 1981, Aaby 1977) have suggested that patriarchal and gerontocratic societies have developed in these circumstances, since control of the labour of women and young men is a key to both survival and accumulation. A study (Nyerges 1992) of the swidden economy of the Susu suggests that in frontier circumstances intensification and resource degradation is a consequence of competition for labour.

Frontier social organization is characterized by local migration, rules of primacy (status derives from being a founder or descendant of a founder), and wealth-in-people (that is, the wealth of the group depends on the number of members). As a result household fission is a characteristic feature, as is a patriarchal and gerontocratic quality to social relations, since the labour of junior males and women is fundamental to male social mobility. Marriages are manipulated so that elders can benefit from (almost perpetual) bride-services, polygamy and marriage to widows with children, whilst junior men have few alternative economic options and are seduced rather than compelled into allegiance to particular elders. Old men plant rice on old fallows (which require substantial labour to clear because of the tree regrowth) using the labour of junior men. Junior men are forced through labour scarcity to farm younger fallows (which are easier to clear) near to the village where the women plant groundnuts (in order to get help from the women to farm their crops). Women farm in their own right (and control the proceeds) on land allocated by husbands and fathers, often land which was used the previous year for rice. Men wish to grow a cash crop of chillies on this second year land, but need women's labour to weed the chillies.

> The women will cooperate in this exploitation of their labor, however, only if the farmer has chosen a relatively dry site, as groundnuts will rot if there is groundwater in the field. This means that in order to intensify production by getting women's cooperation in producing a second-year chili crop, a man must choose rainfed . . . as opposed to rain- and groundwater-fed . . . sites for rice and other intercrops in the first year farm.
>
> (Nyerges 1992: 871)

Farming rice on rain-fed land is more risky and less productive than on groundwater-fed land. Here the farming of very young fallows by junior men, as well as the farming of two-year rotations by elder men with women's labour for the chilli crop, leads to death of the tree roots, suppression of coppice regrowth, grass invasion, annual fires and degradation of the forest canopy. Thus Nyerges argues that 'individual Susu farmers interacting in the context of asymmetrical social relationships create an environment characterised by patterned risk, change, and degradation' (Nyerges 1992: 873). This example also shows the interplay of structure and action – women are structurally weak but find room to manoeuvre in the seasonal demand for their labour, the rigidities of gender divisions of labour and the character of conjugal contracts. But the outcome of these bargaining processes are not necessarily environmentally benign.

What this case shows is that even in a situation of subordination women are able to bargain on the basis of their labour power for concessions which are in their individual interests and not those of the household as a unit. It also shows that resource degradation is not only caused by poverty and that *low* population density can be implicated in environmental degradation.

We need to consider the degree to which environmental conservation, by a wide range of agents, frequently seems to be based upon coercive social relations. Democratic participatory forms of

CECILE JACKSON

development are advocated widely for sustainable development, yet it may be that these are not compatible. The erosion of coercive social relations and increasing individual autonomy have been associated with the breakdown of environmental regulation and the collapse of collective action. The study by Jodha in Rajasthan, India, concluded that land reform, which reduced the power of landlords and improved land access of the poor, was a more significant factor in the decline of common property resources than either commercialization, population pressure or the adoption of tractors. 'In Rajasthan the introduction of land reforms in the 1950s disrupted traditional arrangements that protected and regulated the use of common property resources' (Jodha 1985: 247). This was because 'through levies and penalties on the use of CPRs the . . . landlords exploited the peasants. However as a byproduct of this exploitative mechanism emerged a management system that protected, maintained and regulated the use of common property resources' (Jodha 1985: 254).

A further example of the way in which environmental adaptation and regulation can be based on profound social inequality can be found in the interpretation of caste and conservation by Gadgil and Guha (1992b). They suggest that in the fourth to ninth centuries the Indian subcontinent experienced a resource crunch which led to the crystallization of caste society as 'an elaborate system of the diversified use of living resources that greatly reduced inter-caste competition, and very often ensured that a single caste group had a monopoly over the use of any specific resource from a given locale' (Gadgil and Guha 1992b: 95). The authors unacceptably represent caste groups as 'linked together in a web of mutually supportive relationshps' (ibid.: 93) in a manner which denies inter-caste exploitation and the denigration of lower castes. Their case is better read as one in which powerful social control and hierarchy developed as a means of adapting to resource scarcity and regulating resource use.

If it is the case that environmental conservation is frequently based upon coercive social relations, then how are we to understand the eco-protests of rural women? If the environmental relations of women embody their subordination then why do they, as is claimed, mobilize around the defence of the environment, as in Chipko? At one level it is possible to produce counter-evidence that shows women actively resisting conservation – for example in Cameroon, 1958, when women's protests included their objection to 'a government edict prescribing contour farming to replace the building of ridges vertically along hillsides. Meant to prevent soil erosion, the directive was extremely unpopular amongst women because it is much more difficult and labour-intensive to ridge horizontally on a steep slope' (Diduk 1989: 338; also see Ardener 1975: 38). However, this would

138

be to take at face value the immediate causes of protest. Environmental change has to be seen within the wider context of change. We need to look below the surface of an environmental manifestation to understand the meaning of particular protests. Although Chipko has been widely represented as showing the affinity of women with nature, more recent analyses see it as part of a broader current of peasant protest. Chipko women can be seen as defending a conservative 'moral economy' (Guha 1989a) rather than trees as such, and the Cameroonian women are also seen as essentially resisting the transfer of land from subsistence cultivation to herding and cash-crop farming and defending the traditional order (Diduk 1989: 347). In Maharashtra, Omvedt reports that it was the 1970–3 drought and famine which led women into agitation and protest: 'Women were a majority of workers on the [relief] projects and were reported as the most militant in the demonstrations' (Omvedt 1978: 394). Environmental protests by rural women cannot be disembodied from their livelihood systems, for threatened resources often mean threatened subsistence. This may well mobilize women to protest, but since the moral economy itself is imbued with gender inequality such struggles are not *necessarily* progressive for women, nor, as we point out above, are they driven by environmentalism.

What we have seen in colonial southern Africa, on the other hand, was a struggle against the control of patriarchal elders and the state, expressed through resistance to conservation technologies and planning or via 'everyday forms of resistance' like migration and non-co-operation with marriage transactions (Beinart 1989, Drinkwater 1989, McCracken 1987, Jeater 1989, Lovett 1989). Gender struggles and those to conserve environments are as likely to clash as coincide. Just as the numerical domination of men in, for example, logging rain forest or mining does not mean that women are therefore more environment-friendly, the involvement of women in 'environmental' protests cannot be used as green credentials.

In conclusion, we have found that the assumed and asserted complementarities of women's gender interests and environmental interests derive from the specific (and flawed) conceptualizations of women and nature in different environmentalisms and the absence of gender analysis. Technocentric environmentalism and ecofeminism are unable to see the conflicting interests of women and environments because of gender blindness, ethnocentrism and essentialism, as well as because of a lack of social, historical and material analysis. Linking women with nature is part of a construction of difference, it affirms women as 'other', it prevents more useful gender analysis and has potentially damaging practical consequences. Our analysis has also revealed the limitations of an exclusive focus on poverty-driven environmental

CECILE JACKSON

degradation – environmentally damaging behaviour also results from gender interests and ideologies. Finally, we have questioned the wisdom of assuming that equity in, and democratization of, development is not necessarily compatible with environmental protection, conservation and sustainable use of resources.

If we cannot generalize about gender and environmental relations what approach can be taken? The application of gender analysis to environmental issues would seem a preferable route to understanding the interactions of gender relations and environmental relations. This would include what Bina Agarwal (1991) has termed a 'feminist environmentalism':

> In this conceptualisation, therefore, the link between women and the environment can be seen as structured by a given gender and class (/caste/race) organisation of production, reproduction and distribution. Ideological constructions such as of gender, of nature, and of the relationship between the two, may be seen as (interactively) a part of this structuring, but not the whole of it.
>
> (Agarwal 1991: 11)

Agarwal calls for 'struggles over both *resources* and over *meanings*' (1991: 11, original emphasis). However a gendered environmentalism does not assume subordination of women in all environmental relations and it implies both the addition of greater attention to intrahousehold dimensions of resource relations rather than the intra-class/caste perspective emphasized by Agarwal, and a more action- and agency-oriented perspective.

Such a project would involve many elements of the gender analysis outlined above, both the structural characteristics of gender relations and the understanding of the interactive dynamics of the making and changing of gender relations and meanings as a consequence of individual agency. One area which is an important part of gender analysis, and which reveals the necessity to combine both structure and action, is that of property relations.

Property rights are social relations – that is, they represent relationships between people and people rather than people and things. Men and women experience access to land in profoundly different ways in most rural Third World societies yet the institutions which regulate and enforce property rights (patrilineal descent, legal structures) are not beyond manipulation and influence. Individual actions and agency over time aggregate to change these institutions – for example, Kenyan women and land struggles. In much of Kenya tenurial arrangements mean that women have predominantly secondary rights to land (that is, rights through marriage), whereby wives can be allocated land by their husbands to use whilst a member of his

140

household. In a study of coexisting individual freehold tenure and 'customery' land law in the Murang'a District of Kenya it has been suggested that the latter has provided opportunities to women to struggle for access to land;

> Customary law has been shown to represent 'the responses of living interests' rather than 'the dead hand of tradition' (Chanock 1985: 237). As such, it provides means for legitimation for the more powerful in struggles for control over land and labour. But it also provides some means, albeit limited, whereby women have evolved a discourse of resistance, or counter-power, whether individually, through the manipulation of such practices as the female husband, or collectively by recreating past idioms of social practice and presenting them in terms of contemporary ideology (Harambee).
>
> (MacKenzie 1990: 637)

The plurality of property rights offers a range of options for women to struggle for access to land. Land is only one resource, used for illustrative purposes, and livelihoods are composed of a portfolio of activities and strategies based on a wide range of 'resources', all of which are subjected to gender analysis - for example, environmental relations are also gendered through differential access to labour and to differential knowledges.

The concept of reproduction usefully captures the combination of structure and action, having multiple levels and meanings. Biological reproduction refers to the process of child-bearing and rearing; generational or daily reproduction refers to the maintenance of the domestic group (for example, food production and processing, water collection, etc.); social reproduction involves a range of wider processes whereby societies are reproduced (for example, education socializes children to particular positions in divisions of labour) or changed through the actions of individuals. In all these arenas women and men play different roles, have different rights and responsibilities, different knowledges and expectations. The concept of reproduction is especially useful for understanding gender issues in environmental change because it links household-level divisions of labour into societal processes such as the changing marriage and kinship structures and behaviour, the formula ion and reformulation of norms and values, the trajectories of accumulation and immiserization, the ebb and flow of state policies and interventions.

We have discussed above the central importance of understanding intrahousehold relations. Domestic groups can be seen to operate on the basis of sets of implicit contracts between members, defining areas of rights and responsibilities which guide behaviour and action. These are not immutable, indeed they are contested and struggled over and changed with time, and they may vary with class or other social divisions but they are an important element in mediating the impact of environmental degradation. In addition to gender divisions of labour and responsibilities, we also need to know about gender divisions in access to and control of incomes, both cash and kind, since men and women vary in both the acquisition and disposal of incomes. In the disposal of incomes the differential roles of men and women is an important element in both understanding the incentives towards conservation and the effective ability to participate in projects and programmes.

NOTES

1   There is an inevitable arbitrariness about labelling literatures – some views carry a mixture of approaches, some edited collections bear papers reflecting several perspectives – but I would consider the following to represent ecofeminist discourse: Cox (1992), Merchant (1982, 1992), Plumwood (1986, 1988, 1992), Ruether (1979), Shiva (1989), Warren (1987, 1990), Women's Environmental Network (1989), Gray (1981), Eastlea (1981), King (1989), Griffin (1978).

   What I call the women and environment literature is influenced by ecofeminist ideas, which emphasize the 'natural' affinity of women with their environments, even if these are not explicitly recognized, for example: Sontheimer (1991), Dankelman and Davidson (1989), Rodda (1991), Fortmann (1986), Munyakho (1985), Special Issue of *Development* 1902, IUCN (1987). Much of this is a large grey literature not formally published.

   Finally, one can perhaps discern minor alternative approaches, labelled by Agarwal 'feminist environmentalism' (for example, Agarwal 1991 and DAWN 1987), but also gender analyses of environmental change such as Leach (1991, 1992), Cecelski (1984), Kelkar and Nathan (1991).

2   Gender analysis is a composite of a number of disciplines, a collective endeavour, perhaps typified by Stamp (1989), Moore (1988), Pearson (1992), Elson and Pearson (1981), Young *et al.* (1981), Mackintosh (1989), Agarwal (1986b), Whitehead (1981, 1990), Parpart and Staudt (1989), Molyneux (1985), Sharma (1980), DAWN (1987), Rosaldo and Lamphere (1974), Guyer (1980), Kandiyoti (1985).

   Henrietta Moore uses the term 'feminist anthropology' for a similar framework, but the divergence from feminism and the interdisciplinary character of contributors to the gender analysis make this now something of a misnomer.

   The differences between feminism and gender analysis are variable (greatest for radical feminsm, least for socialist feminism) and stem largely from their different objectives – feminism aims for social change, gender analysis is a framework for understanding social relations. Feminism and gender analysis have a potentially symbiotic, rather than an oppositional, relationship; gender analysis informing feminism, and feminism generating social change.

3   The variations by class and region are great. However, in Bangladesh, for example, the evaluation of the gendered consequences of the introduction of mechanized rice milling, which was part of the green revolution, has shifted from a negative picture of massive job losses for poor women rice processors (Harriss 1979) to a more favourable one in which women of all classes have come to benefit from the technology through replacement of unpaid domestic work and new work opportunities (White 1992: 75-7).

4   'Part of the problem of characterising eco-feminist practice and thought is that both seem to be different in different regions or countries' (Faber and O'Connor 1989). There seems to be a dominant element of radical ecofeminism in US environmental groups (on the west coast in particular), with a strong spiritual perspective. The socialist ecofeminists are more characteristic of Europe and Australia. Both exhibit the elements we discuss here to differing extents. In a recent book Carolyn Merchant also identifies liberal, cultural and social ecofeminism (1992). The characteristics ascribed to ecofeminism here are a core of shared positions, given different emphasis in each strand of ecofeminism.

5   Speciesism is a key concept in ecocentric discourse and has been defined as 'a prejudice or attitude of bias toward the interests of members of one's own species and against those of other species' (Ryder 1974, quoted in Eckersley 1992: 43).

6   Female infanticide killed more people than *meriah*. In one area in 1848, only 20 girls were enumerated for the 231 boys under 10 years listed (Leigh Stutchbury 1982: 53).

7   Status hypergamy is a rule which requires women to marry into the same or higher status category than themselves. The result is that at any one level there are 'surplus' women because men of that level can demand very high dowries of lower status women. Since the condition of spinsterhood is unacceptable, the supply of women is limited through female infanticide as the response of individul men and women who, because of the cost of dowry, feel they cannot afford to allow more than one girl to survive.

REFERENCES

Aaby, P. (1977) 'Engels and women', *Critique of Anthropology* 3(9/10): 25-53.

Agarwal, B. (1986a) *Cold Hearths and Barren Slopes. The Woodfuel Crisis in the Third World*, London: Zed Books.

—— (1986b) 'Women, poverty and agricultural growth in India', *Journal of Peasant Studies* 13(4): 628-42.

—— (1991) 'Engendering the environment debate: lessons from the Indian subcontinent', CASID Distinguished Speaker Series No 8, Center for Advanced Study of International Development, Michigan State University, East Lansing, Michigan.

Apffel Marglin, F. (1992) 'Women's blood: challenging the discourse of development', *The Ecologist* 22(1): 22-32.

Ardener, E. (1975) 'Belief and the problem of women', pp. 1-18 in S. Ardener (ed.), *Perceiving Women*, London: Malaby.

Beinart W. (1989) 'Introduction: the politics of colonial conservation', *Journal of Southern African Studies* 15(2): 143-62.

Bernstein, H. (1990) 'Taking the part of peasants?', pp. 69–79 in H. Bernstein *et al.* (eds), *The Food Question. Profits versus People?*, London: Earthscan Publications.

Biehl, J. (1988) 'Ecofeminism and deep ecology: unresolvable conflict?', *Our Generation* 19(2): 19–32.

Billig, M. (1991) 'The marriage squeeze on high-caste Rajasthani women', *The Journal of Asian Studies* 50(2): 341–60.

Blaikie, P. (1985) *The Political Economy of Soil Erosion in Developing Countries*, London: Longman.

Bradley, P. (1991) *Women, Woodfuel and Woodlots. Volume 1: The Foundations of a Fuelwood Strategy for East Africa*, London: Macmillan.

Brown, P. and Jordanova, L. (1982) 'Oppressive dichotomies', pp. 189–399 in Open University, *The Changing Experience of Women*, Oxford: Martin Robertson.

Bruce, J. (1989) 'Homes divided', *World Development* 17(7): 979–1136.

Bunch, C. and Carrillo, R. (1990) 'Feminist perspectives on women in development', pp. 70–82 in I. Tinker (ed.), *Persistent Inequalities*, New York: Oxford University Press.

Burman, B.K. Roy (1992) 'Homage to Earth', in G. Sen (ed.) *Indigenous Vision: Peoples of India Attitudes to the Environment*, New Delhi: Sage.

Cecelski, E. (1984) *The Rural Energy Crisis, Women's Work and Family Welfare: Perspectives and Approaches to Action*, World Employment Programme Research Working Paper no. WEP10/WP35, Geneva: International Labour Office.

Cliffe, L. and Moorsom, R. (1979) 'Rural class formation and ecological collapse in Botswana', *Review of African Political Economy* 15–16: 35–52.

Conway, G. and Barbier, E. (1990) *After the Green Revolution: Sustainable Agriculture for Development*, London: Earthscan.

Cox, C. (1992) 'Ecofeminsm', pp. 282–3 in G. Kirkup and L. Keller (eds), *Inventing Women: Science, Technology and Gender*, London: Open University and Polity Press.

Dankelman, I. and Davidson, J. (eds) (1989) *Women and Environment in the Third World: Alliance for the Future*, London: Earthscan Publications.

Davidson, J. and Myers, D. (with M. Chakroborty) (1992) *No Time to Waste. Poverty and the Global Environment*, Oxford: Oxfam.

DAWN (1987) *Development, Crises and Alternative Visions: Third World Women's Perspectives*, New York: Monthly Review Press.

de Beauvoir, S. (1988) *The Second Sex*, London: Picador.

Development: Journal of SID (1992) Special Issue no. 2.

Diduk, S. (1989) 'Women's agricultural production and political action in the Cameroon grassfields', *Africa* 59(3): 338–55.

Dixon-Mueller, R. (1985) *Women's Work in Third World Agriculture*, Women, Work and Development no. 9, Geneva: International Labour Office.

Douglas, M. (1966) *Purity and Danger*, London: Routledge & Kegan Paul.

—— (1973) 'Self evidence', The Henry Myers lecture 1972, *Proceedings of the Royal Anthropological Institute of Great Britain and Ireland*, pp. 27–44, London: Royal Anthropological Institute.

—— (1991) *Implicit Meanings. Essays in Anthropology*, London: Routledge.

Drinkwater, M. (1989) 'Technical development and peasant impoverishment: land use policy in Zimbabwe's Midlands Province', *Journal of Southern African Studies* 15(3): 287–305.

Eastlea, B. (1981) *Science and Sexual Oppression: Patriarchy's Confrontation with Women and Nature*, London: Weidenfeld & Nicolson.

Eckersley, R. (1992) *Environmentalism and Political Theory: Toward an Ecocentric Approach*, London: University College London Press.

Ellen, R. (1982) *Environment, Subsistence and System. The Ecology of Small Scale Social Formations*, Cambridge: Cambridge University Press.

Elson, D. and Pearson, D. (1981) 'Nimble fingers make cheap workers: an analysis of women's employment in Third World export manufacturing', *Feminist Review* 7: 87–107.

Evans, A. (1988) 'Gender relations and technological change: the need for an integrative frame of analysis', in S. Poats, M. Schmink and A. Spring (eds), *Gender Issues in Farming Systems Research and Extension*, Boulder and London: Westview Press.

Faber, D. and O'Connor, J. (1989) 'Ecofeminism/ecosocialism', *Capitalism, Nature, Socialism* 2(1): 137–40.

Fernandes, W. and Nenon, G. (1987) *Tribal Women and Forest Economy: Deforestation, Exploitation and Status Change*, New Delhi: Indian Social Institute.

Folbre, N. (1986) 'Hearts and spades: paradigms of household economics', *World Development* 14(2): 245–55

Fortmann, L. (1986) 'Women in subsistence forestry', *Journal Forestry* 84(7): 39–42.

Gadgil, M. and Guha, R. (1992) 'Interpreting Indian environmentalism', Paper to the Conference on the Social Dimensions of Environment and Sustainable Development, United Nations Research Institute for Social Development, Valetta, Malta, 22–25 April.

—— —— (1992b) *This Fissured Land. An Ecological History of India*, Delhi: Oxford University Press.

Gillison, G. (1980) 'Images of nature in Gimi thought', pp. 143–73 in C. MacCormack and M. Strathern (eds), *Nature, Culture and Gender*, London: Cambridge University Press.

Goody, J. (ed.) (1971) *The Developmental Cycle in Domestic Groups*, Cambridge: Cambridge University Press.

Gray, E. Dodson (1981) *Green Paradise Lost*, Wellesley, Mass.: Roundtable.

Griffin, S. (1978) *Woman and Nature: The Roaring Inside Her*, New York: Harper & Row.

Guha, R. (1989a) *The Unquiet Woods: Ecological Change and Peasant Resistance in the Himalaya*, Dehli: Oxford University Press.

—— (1989b) 'Radical American environmentalism and wilderness preservation: a Third World critique', *Environmental Ethics* 11(1): 71–83.

Guyer, J. (1980) 'Food, cocoa and the division of labour by sex in two West African Societies', *Comparative Studies in History and Society* 22(3): 355–73.

—— (1981) 'Household and community in African Studies', *African Studies Review* 24 (2/3): 87–138.

Harriss, B. (1979) 'Post harvest rice processing systems in rural Bangladesh: technology, economics and employment', *Bangladesh Journal of Agricultural Economics* 2(1): 64–72.

Hartmann, B. and Standing, H. (1989) *The Poverty of Population Control: Family Planning and Health Policy in Bangladesh*, London: Bangladesh International Action Group.

Hoskins, M. (1983) *Rural Women, Forest Outputs and Forstry Projects*, Rome: FAO.

IUCN (International Union for Conservation of Nature) (1987) *Women and the World Conservation Strategy*, Report on the first Strategy Workshop held at IUCN, Switzerland.

Jackson, C. (1993) 'Questioning synergism: win–win with women in population and environment policies?', forthcoming in *Journal of International Development*.

Jeater, D. (1989) 'The closing gap: African women and European morality in Southern Rhodesia 1915–30', Paper for the African History Dept Seminar, School of Oriental and African Studies, University of London, January.

Jodha, N.S. (1985) 'Population growth and the decline of common property resources in Rajasthan, India', *Population and Development Review* 11(2): 247–64.

Joekes, S. and Pointing, J. (1991) 'Women in pastoral societies in East and West Africa', Dryland Network Programme Issues Paper no. 28, IIED.

Jones, K. and Jonasdottir, A. (eds) (1985) *The Political Interests of Gender*, London: Sage.

Kandiyoti, D. (1985) *Women in Rural Production Systems: Problems and Policies*, Paris: UNESCO.

Kelkar, G. and Nathan, D. (1991) *Gender and Tribe: Women, Land and Forests*, New Dehli, Kali for Women.

King, Y. (19897 'The ecology of feminism and the feminism of ecology', in J. Plant (ed.), *Healing the Wounds: The Promise of Ecofeminism*, Philadelphia: New Society Publishers.

Kitching, G. (1982) *Development and Underdevelopment in Historical Perspective*, Open University Set Book, London: Methuen.

Leach, M. (1991) 'Engendering environments: understanding natural resource management in the West African forest zone', *Institute of Development Studies Bulletin* 22.

—— (1992) 'Gender and the environment: traps and opportunities', *Development in Practice* 2(1): 12–22.

—— and Mearns, R. (n.d.) 'Population and environment in developing countries: an overview study', Final report ESRC Global Environmental Change Research Programme.

Leigh Stutchbury, E. (1982) 'Blood, fire and mediation: human sacrifice and widow burning in nineteenth century India', pp. 21–75 in M. Allen and S. Mukherjee (eds), *Women in India and Nepal*, Australian National University Monographs on South Asia no. 8, Canberra: Australian National University.

Lipton, M. and Longhurst, R. (1989) *New Seeds and Poor People*, London: Unwin Hyman.

Lovett, M. (1989) 'Gender relations, class formation and the colonial state in Africa', pp. 47–69 in J. Parpart and K. Staudt (eds), *Women and the State in Africa*, Colorado: Lynne Rienner Publishers.

Low, A. (1986) *Agricultural Development in Southern Africa: Farm Household Theory and the Food Crisis*, London: James Currey.

MacCormack, C. (1980) 'Nature, culture and gender: a critique', pp. 1–24 in C. MacCormack and M.Strathern (eds), *Nature, Culture and Gender*, London: Cambridge University Press.

McCracken, J. (1987) 'Colonialism, capitalism and ecological crisis in Malawi: a reassessment' pp. 63–77 in D. Anderson and R. Grove (eds),

Conservation in Africa: Peoples, Policies and Practice, Cambridge: Cambridge University Press.

McGregor, J. (1991) 'Woodland resources: ecology policy and ideology. An historical case study of woodland use in Shurugwi Communal Area, Zimbabwe', Unpublished Ph.D. thesis, Loughborough University.

MacKenzie, F. (1990) 'Gender and land rights in Murang'a District, Kenya', *Journal of Peasant Studies* 17: 609–43.

MacKenzie, J. (1987) 'Chivalry, social Darwinism and ritualised killing: the hunting ethos in Central Africa up to 1914', in D. Anderson and R. Grove (eds), *Conservation in Africa: Peoples, Policies and Practice*, Cambridge: Cambridge University Press.

Mackintosh, M. (1989) *Gender, Class and Rural Transition: Agribusiness and the Food Crisis in Senegal*, London: Zed Press.

Mahapatra, S. (1992) 'Rites of propitiation in tribal societies', pp. 63–74 in G. Sen (ed.), *Indigenous Vision: Peoples of India Attitude to the Environment*, New Delhi: Sage.

Meillasoux, C. (1981) *Maidens, Meal and Money*, Cambridge: Cambridge University Press.

Merchant, C. (1982) *The Death of Nature: Women, Ecology and the Scientific Revolution*, London: Wildwood House.

—— (1992) *Radical Ecology: The Search for a Livable World*, London: Routledge.

Molyneux, M (1985) 'Mobilisation without emancipation? Women's interests, the state, and revolution in Nicaragua', *Feminist Studies* 11(2): 227–54.

Moore, H. (1988) *Feminism and Anthropology*, London: Polity.

Moser, C. (1989) 'Gender planning in the Third World: meeting practical and strategic gender needs', *World Development* 17(11): 1799–1825.

Munyakho, D.K. (ed.) (1985) *Women and the Environmental Crisis: A Report of the Proceedings of the Workshops on Women, Environment and Development*, 10–20 July, Nairobi Kenya NGO Forum, Nairobi: Environment Liaison Centre.

Murphy, P. (1988) 'Sex-typing the planet: Gaia imagery and the problem of subverting patriarchy', *Environmental Ethics* 10(2): 155–68.

Netting, R. Wilk, R. and Arnould, E. (eds) (1984) *Households. Comparative and Historical Studies of the Domestic Group*, Berkeley and Los Angeles: University of California Press.

Newby, H. (1979) *Green and Pleasant Land? Social Change in Rural England*, London: Hutchinson.

Nyerges, A. Endre (1992) 'The ecology of wealth-in-people: agriculture, settlement and society on the perpetual frontier', *American Anthropologist* 94(4): 860–81.

Omvedt, G. (1978) 'Women and rural revolt in India', *Journal of Peasant Studies* 5(3): 370–403.

O'Riordan, T. (1981) *Environmentalism*, London: Pion.

Ortner, S. (1974) 'Is female to male as nature is to culture?', pp. 67–88 in M. Rosaldo and L. Lamphere (eds), *Woman, Culture and Society*, Stanford, Calif. Stanford University Press.

Parpart, J. and Staudt, K. (eds) (1989) *Women and the State in Africa*, Hartford, Conn.: Westview Press.

Pearson, R. (1992) 'Gender matters in development', pp. 291–312 in T. Allen and A. Thomas (eds), *Poverty and Development in the 1990's* Oxford: Open University and Oxford University Press.

Pepper, D. (1986) *The Roots of Modern Environmentalism*, London: Routledge.

Plumwood, V. (1986) 'Ecofeminism: an overview and discussion of positions', *Australian Journal of Philosophy* 64: 120–38.

—— (1988) 'Woman, humanity and nature', *Radical Philosophy* 48: 16–24.

—— (1992) 'Beyond the dualistic assumptions of women, men and nature', *Ecologist* 22(1): 8–13.

Poats, S., Schmink, M. and Spring, A. (eds) (1988) *Gender Issues in Farming Systems Research and Extension*, Boulder and London: Westview Press.

Rao, B. (1991) 'Dominant constructions of women and nature in social science literature', *Capitalism, Nature, Socialism*, Pamphlet 2, New York, Guilford Publications.

Reijntjes, C., Haverkort, B. and Waters-Bayer, A. (1992) *Farming for the Future. An Introduction to Low-external-input and Sustainable Agriculture*, London: ILEIA and Macmillan.

Rodda, A. (1991) *Women and the Environment*, London: Zed Books.

Rosaldo, M. and Lamphere, L. (eds) (1974) *Woman, Culture and Society*, Stanford, Calif.: Stanford University Press.

Rubin, G. (1975) 'The traffic in women: notes on the political economy of sex', pp. 157–210 in R. Reiter (ed.) *Towards an Anthropology of Women*, New York: Monthly Review Press.

Ruether, R. (1979) *New Woman, New Earth*, New York: Seabury Press.

Ryder, R. (1974) *Speciesism: The Ethics of Vivisection*, Edinburgh: Scottish Society for the Prevention of Vivisection.

Sayers, J. (1982) *Biological Politics: Feminist and Anti-feminist Perspectives*, London: Tavistock Publications.

Scarce, R. (1990) *Ecowarriors: Understanding the Radical Environmental Movement*, Chicago: Noble Press.

Schmidt, A. (1971) *The Concept of Nature in Marx*, London: New Left Books.

Sen, A. (1987) 'Gender and cooperative conflicts', Helsinki: WIDER.

Sen, G. (1985) 'Women workers and the green revolution', pp. 29–65 in L. Beneria (ed.), *Women and Development: The Sexual Division of Labour in Rural Societies*, New York: Praeger.

Sharma, U. (1980) *Women, Work and Property in North-west India*, London: Tavistock.

Shiva, V. (1989) *Staying Alive: Women, Ecology and Development*, London: Zed Books.

Sontheimer, S. (ed.) (1991) *Women and the Environment: a Reader*, London: Earthscan Publications.

Stamp, P. (1989) *Technology, Gender and Power in Africa*, Canada: International Development Research Centre.

Thomas-Slayter, B. (1992) 'Politics, class and gender in African resource management: the case of rural Kenya', *Economic Development and Cultural Change* 40(4): 809–27.

Tshibaka, T. (1992) *Labor in the Rural Household Economy of the Zairian Basin*, Research Report 90, Washington: International Food Policy Research Institute.

van Beek, W. and Banga, P. (1992) 'The Dogon and their trees', pp. 57-75 in E. Croll and D. Parkin (eds), *Bush Base: Forest Farm. Culture, Environment and Development*, London: Routledge.

van den Breemer, J. (1992) 'Ideas and usage: environment in Aouan society, Ivory Coast', pp. 97-109 in E. Croll and D. Parkin (eds), *Bush Base; Forest Farm. Culture, Environment and Development*, London: Routledge.

van Leynseele, P. (1979) 'Ecological stability and intensive fish production: the case of the Libinza people of the Middle Ngiri (Zaire)', pp. 167-84 in P. Burnham and R. Ellen (eds), *Social and Ecological Systems*, London: Academic Press.

Venkateswaran, S. (1992) *Living on the Edge: Women, Environment and Development*, Delhi, Friedrich Ebert Stiftung.

Warren, K. (1987) 'Feminism and ecology: making the connections', *Environmental Ethics* 9(1): 3-20.

—— (1990) 'The power and the promise of ecological feminism', *Environmental Ethics* 12: 125-46.

White, L. (1967) 'The historical roots of our ecologic crisis', *Science* 155: 1203-7.

White, S. (1992) *Arguing with the Crocodile: Gender and Class in Bangladesh*, London: Zed Press.

Whitehead, A. (1979) 'Some preliminary notes on the subordination of women', *Bulletin of the Institute of Development Studies* 10(3): 10-13.

—— (1981) 'A conceptual framework for the analysis of the effects of technological change on rural women', WEP Working paper no. 79, Geneva: ILO.

—— (1990) 'Rural women and food production in sub-Saharan Africa', pp. 425-74 in J. Dreze and A. Sen (eds), *The Political Economy of Hunger. Volume 1: Entitlement and Well-being*, Oxford: Clarendon Press.

Wilson, K. (1989) 'Trees in fields in southern Zimbabwe', *Journal of Southern African Studies* 15(2): 369-83.

Wipper, A. (1989) 'Kikuyu women and the Harry Thuku disturbances: some uniformities of female militancy', *Africa* 59(3): 300-36.

Women's Environmental Nework (1989) *Women, Environment, Development. Seminar Report*, London: WEN.

World Bank (1992) *World Development Report*, New York: Oxford University Press.

Young, K. and Harris, O. (eds) (1981) *Of Marriage and the Market*, London: CSE Books.

7

SOCIAL MOVEMENTS AND ENVIRONMENTAL CHANGE

Steven Yearley

INTRODUCTION

Last year Howard Newby attacked sociology, and British sociology in particular, for making only a 'slender' contribution to the study of the environment (1991). In this chapter I aim to expand, perhaps by only a little, this contribution by examining what can be gained by using the sociology of social movements to analyse the characteristic features of the environmental movement. The analytic rationale for this chapter arises from the attempt to specify which features are common to other social movements and which are not. The indication of the movement's sociological peculiarities will be of particular value. The practical rationale arises because the environmental movement and movement organizations have been highly influential in bringing about changes in public attitude and commercial behaviour, in influencing public policy and in motivating people to engage in practical projects in the environment's perceived interest (for a breathless catalogue see Pearce 1991: pp. ix–xi). Sociology needs to develop its apparatus for describing these groups and this movement.

Judging from Newby's comments one might suppose that the researcher would be hard pressed to find sociologists writing about environmental groups and ecological issues. On the contrary, many authors have already sought to include Greens within the compass of their analysis. And these authors represented a wide range of sociological persuasions. At one extreme, leading theorists such as Giddens have tried to incorporate the green movement within their large-scale interpretations. Giddens, exploring what he terms 'high' modernity, emphasizes the ubiquity of risk and danger in the modern world. Among the global risks which shape contemporary culture and politics, he gives prominent position to 'ecological calamity and uncontainable population explosion' (Giddens 1990: 125). For Giddens, green movements constitute one of the four types of social movements which represent a response to risks in the principal 'institutional dimensions' of modernity. He sees 'clear countertrends, partly expressed through ecological movements', challenging the logic of relentless, institutionalized technological innovation (Giddens 1990: 170). In his work on the 'risk society' Beck goes even further by nominating the Greens as a prime example of what he terms reflexive modernization, the destructive application of modernist principles to themselves (Beck 1992: 156). By using scientific methods to criticize the harm caused by our scientific civilization – in particular by using scientific reasoning to calibrate the uncertainty and risk associated with agrochemicals, nuclear power and additives – Greens publicly expose the fix into which 'late modernity' has got itself.

At a rather less abstract level Melucci, who regards social movements as in many respects *the* characteristic feature of complex modern societies, also places considerable emphasis on green movements. For him, 'the peace and ecological movements [are key] precisely because they are testaments to the fragile and potentially self-destructive connections between humanity and the wider universe' (Melucci 1989: 6). Lastly, environmental groups have figured in many case-studies of the formation and development of social movements. This is true both within the US tradition, which concentrates on the formal and organizational characteristics of social movements (see Zald and McCarthy 1987: 179, on the Sierra Club and Friends of the Earth), and the European mould, which seeks to relate social movements to political and class interests (for example, see Touraine *et al*. 1983 on the French anti-nuclear movement).

Accordingly, my task in this chapter is in no sense to initiate sociological interest in environmental social movements but – with luck – to clarify and customize existing work. I shall attempt this under four headings: the theoretical interpretation of environmental social movements, the special characteristics of green social movements, the peculiarities engendered by the role of scientific authority within the green movement, and the nature of environmental movements outside the core of industrialized, market economies – particularly in non-Western, non-industrialized countries.

HOW USEFUL ARE THEORETICAL INTERPRETATIONS OF THE GREEN MOVEMENT?

Virtually every commentator on the sociology of social movements makes use of the distinction, which I have already mentioned, between the US and the European interpretive traditions. On the European side are those who assess social movements primarily in relation to their perceived capacity for major social transformation. While there may be many campaigning groups or sets of people

151

agitating for social change or alterations in public policy, only some of these qualify for the appellation 'new social movements'.

Such movements are described as 'new' essentially because they cannot be seen as directly deriving from the 'old' labourist movements. In other words, early social analysts had assumed that social movements were essentially expressions of the accepted, 'major' political forces; they were seen as embodiments of workers' attempts to further their economic and political interests and therefore derivative in some way from class politics. However, the new social movements did not seem to reflect this dynamic. The student movement, civil rights and peace movements and even the women's movement seemed to defy earlier assumptions, drawing support from across the classes and pursuing political ends which could not be boiled down to obvious class objectives. Often, participation in, or support for, such movements is more closely correlated with educational experience than with class position.

Although not unified by presumed class alignment, the commonly cited new social movements share a progressive political orientation. For example, the women's movement, civil rights groups and radical environmentalists can all be said to be extending our conception of rights – to women, to formerly excluded minorities and to the natural world. For this reason the new social movements are seen as the natural allies of the 'progressive' left. And some analysts, despairing that the working classes are ever going to bring about social transformation, have transferred their hopes for radical reform on to these movements. In this way of thinking, the social movements can take on the historic role usually ascribed to classes. As Touraine expresses this idea:

> Social movements are neither accidents nor factors of change: they are the collective action of actors at the highest level – the class actors – fighting for the social control of historicity, i.e. *control of the great cultural orientations* [of society].
>
> (Touraine 1981: 26; emphasis added)

According to such a view, the only movements which qualify for the title 'social movements' are ones which take over the historic agenda of social classes and which strive to control the 'great cultural orientations' of society. The problem then is deciding which movements qualify. Civil rights movements might well, but how about the youth movement? What separates a 'great cultural orientation' from a less-than-great one?

The leading alternative approach adopts a much more disinterested interpretation of a social movement. For this approach a social movement is defined in terms of organizational characteristics. Social movements are instances of collective behaviour which are more organized than protesting crowds or mobs, less formalized than political parties and more concerted than simple social trends. Obviously, the boundaries around this phenomenon are not clear-cut, but a reasonable amount of agreement can be reached about what is a movement, what is a party and what is a crowd.

Adherents of this view note a long-term trend. Early on, social movements tended to be made up of massed participants. More recently the role of professionalized cadres of movement organizers has markedly increased. By the 1980s the apparently vociferous social movements were actually highly efficient social movement organizations (SMOs), sometimes followed by few spontaneously active supporters. Analysts drew attention to the importance of organizational and technical innovations such as the pre-printed letter addressed to politicians or senior officials and – especially popular in the USA – records of the voting record of state and federal politicians. A politician who has once voted for, or made favourable remarks about, say abortion or gun control was likely to have his or her record resurrected by right-wing pressure groups. Using these methods, professional campaigners could readily create the impression of a large social movement.

Such techniques, which represent a rationalization of social movement practices, became widespread, adopted by moral crusaders, wildlife conservationists and anti-Euro-federalists. Analysts within this tradition are accordingly struck more by the similarity of groups, irrespective of their ideological affiliation, than by differences between those directed at 'cultural transformation' and those dealing with more sectional, sectarian or even trivial concerns.

However, the US-style approach, which emphasizes the 'moral entrepreneurship' of figures within social movement organizations, suffers one clear drawback. It seems unable to distinguish between effective lobbyists or pressure groups and broader social movements, where the latter offer – at least in principle – the potential for mass participation in some form. The peace movement would be able to take advantage of the most sophisticated forms of direct mailing and to welcome payment by credit card whilst practising canny lobbying of industry and government; at the same time its mass membership was central to its character. It is difficult to conceive a similar situation developing for a group campaigning on the arts. The social attributes of many supporters might be similar, the methods adopted by the organizations' professional staffs closely related, but one cause has been associated with a participatory movement and the other has not. Approaches which focus almost exclusively on organizational characteristics tend not to illuminate the distinction between 'promotional' pressure groups (groups founded to advance a principle or cause) and those SMOs associated with a broad movement. Environmentalism certainly qualifies as a broad movement; even in Britain the movement is estimated to have around 4 million supporters (McCormick 1991: 152). Accordingly, we need to ensure that this distinction is explicitly analysed and accounted for.

In summary, the nominalist, US approach tends to be too undiscriminating, while the European analyst's view is inclined to be excessively restrictive. But this overview of the two competing traditions at least allows us to identify key descriptive elements which, I suggest, characterize a social movement: it has a large-scale membership and a promotional character; modern social movements will tend to adopt sophisticated administrative apparatuses.

The question then becomes, how can the analyst identify and account for this kind of social phenomenon? It seems to me that there have been four principal responses. First, there is the 'European' response which tends to be Hegelian. That is, 'true' social movements are viewed in an historicist way, with the assumption that they articulate the interest of some developing historical actor. As I mentioned above, this impression is easy to reinforce since many social movements share a 'progressive' orientation. But there appears to be no non-arbitrary way of separating the 'true' from other movements, other movements which possess all the characteristics of the 'true' ones apart from a progressive political basis. For example, aside from their profoundly anti-universalistic outlook, nationalist movements would seem to fit the bill perfectly well. Nothing, apart from the analysts' prior assumption about the political orientaion of 'true' movements, could disqualify them from movement status.

A second approach to a systematic classification is illustrated by Giddens's identification of the leading 'institutional dimensions of modernity'. For Giddens (1990: 59) these are capital accumulation, surveillance, military power and industrialism (the last of these representing the 'transformation of nature'). Giddens's suggestion appears to be that, having identified these axes, one can determine whether social movements have arisen in response to each. A plausible enough case can be made for saying that they have: respectively, the labour movement, democracy/free speech movements, the peace movement and the ecology movement. If it could be established that these are indeed the principal axes of advanced societies, then there would be some intuitive support for the idea that the social movements crystallized around them are of special relevance. However, as Giddens himself notes (1990: 162), the 'objectives of feminist movements . . . crosscut the institutional dimensions of modernity'. The same kinds of difficulty confront analogous, Habermasian analyses which attempt to categorize types of social movement. In short, the specification of the essential characteristics of modern capitalism has not been consensually achieved to we cannot 'check off' the movements against the corresponding features of capitalist society. Nor, in any case, does the present variety of social movements map readily on to the leading theorists' anatomies of present-day capitalism.

A third theoretically based attempt to put forward systematic grounds for separating social movements from 'mere' pressure groups and other instances of collective behaviour has been made by Eyerman and Jamison. They have recently argued that it is central to see that social movements are engaged in what they term 'cognitive praxis'; in other words, social movements are characteristically producers of innovative knowledge claims. They suggest that 'a social movement is not one organization of one special interest group. It is more like a cognitive territory, a new conceptual space' (Eyerman and Jamison 1991: 55). Furthermore, they indicate that their 'approach tends to limit the number of social movements to those especially "significant" movements which redefine history, which carry the historical "projects" that have normally been attributed to social classes' (ibid.: 56).

These authors, centrally involved in a three-country study of the environmental movement (Jamison *et al*. 1991), are quite correct to stress the fact that social movements commonly advance new knowledge claims and that movement intellectuals can play a decisive role. As I shall argue when I come on to the part played by scientific authority in the green movement, the fact that Greens' arguments depend on science has had strong sociological implications for the movement. But it appears premature to make cognitive praxis a defining characteristic of 'true' social movements. This is because it is easy to see politically significant social movements featuring little cognitive innovation (the moral majority, nationalism), and also because there are abortive instances of cognitive praxis (assorted utopias, even eugenics). Cognitive praxis alone does not a movement make, a fact which is demonstrated by the way in which Eyerman and Jamison end up fleshing out their theory in an *ad hoc* fashion, asserting that 'only those [social problems] that strike a fundamental chord, that touch basic tensions in a society have the potential for generating a social movement' (Eyerman and Jamison 1991: 56). In the end, even for Eyerman and Jamison, cognitive praxis is not the final explanation for social movement formation. There is, no doubt, a strong association between social movements and cognitive innovation. But it seems unduly restrictive to make such innovation a defining characteristic.

This, I suggest, leaves us with an essentially descriptive definition of social movements. Social movements are to be known by their

155

organizational forms; although they often engage in cognitive praxis, they need not do so. There is, in other words, no single criterion allowing us to explain which pressure groups can be associated with a large-scale social movement and which cannot; the reasons for mass participation have to be analysed on a case-by-case basis.

While apparently lacking the theoretical power of the competing interpretations, this approach has the benefit that – since it lacks one single, key element – it draws attention away from the supposed unitary character of movements. Thus, if we examine the array of environmental SMOs in Britain we see that it is difficult to speak of *a* movement – in fact, the term 'social movement industry' employed by US analysts is highly apt. Although the leading SMOs collaborate over specific campaigns and share supporters, and although there are frequent meetings between the different professional cadres, there is competition between them: competition for the highest profile campaign topics, competition for acceptable and wealthy backers, competition for news coverage – particularly in recessionary times – competition for members and their money. Given this atmosphere of competition, Zald and McCarthy (1987: 179) record their surprise that little attention has been paid to 'inter-SMO' relations. Drawing parallels with the behaviour of firms and formal organizations, they have formulated fourteen generalizations about SMOs' behaviour in competitive situations. In the light of such competition within a movement, and of the fact that such competition is largely played out between SMOs (professionalized organizations keen to retain staff numbers, with staff members obviously keen to hang on to their own jobs), any treatment of the whole movement as *an actor* appears extremely superficial. The outcome of the competition between SMOs will shape the future direction of the movement; such micro-level features will, for example, influence the 'cognitive praxis' of the movement.

A second reason for being wary about describing the environmental lobby in Britain in anything like the Tourainean sense is that the groups which are both radical and effective tend to be centralized. Greenpeace is famously so, leading Allen, a campaigning journalist on waste incineration, to complain that:

in Britain Greenpeace is very definitely bureaucratic and was seduced by the establishment fairly quickly. From a small grouping at the beginning of the eighties, Greenpeace displays all the trappings of a multinational company or a civil service department.

(Allen 1992: 223)

How rapid a seduction has to be to qualify as 'fairly quick' is unclear to me, and – in any case – Greenpeace continues to engage in direct

156

actions, but Allen's apparent resentment stems from the alienation sometimes felt by community campaigners in the face of profession-alized and wealthy campaign organizations. McCormick notes that through the 1980s Greenpeace became 'less confrontational, and more inclined to use the same tactics of lobbying and discreet political influence once reserved by the more conservative groups' (McCor-mick 1991: 158).

Friends of the Earth (FoE) has attempted to institutionalize an arrangement for co-operation between a centralized London-based staff and its local groups. The latter can select their own campaign targets but are bound by a licence agreement with FoE which prevents them from acting in FoE's own name. Local groups have a form of parlia-ment at the annual meeting but cannot require the board to change its policy. The headquarters staff have a highly professionalized ethos (McCormick 1991: 117–18). The structure lessens but does not remove tensions between the core and the regionally active members. Even FoE, which has tried to build in a mechanism for sustained member participation, cannot be viewed as the expression of the movement.

Finally, in relation to the nominalist, descriptive approach, it should be noted that its exponents are not denied access to explanation at the level of macroscopic social change. Thus, Berger has argued for the relevance of the rise of the 'knowledge class', a 'new middle class . . . of people whose occupations deal with the production and distribution of symbolic knowledge' (Berger 1987: 66). These are the intellectuals and service workers whose knowledge is not generally directed towards material production; they work in education, counselling and communications and in the 'bureaucratic agencies planning for the putative nonmaterial needs of the society (from radical amity to geriatric recreation)' (Berger 1987: 67). This knowledge class tends, in Berger's view, to be antagonistic to the core values of capitalism, a fact which he attributes chiefly to two factors: the interest this class has in having 'privilege based on educational credentials'; and the interest this class has in expanding the role of the welfare state where its members find work.'[The] knowledge class has an interest in the distributive machinery of government, as against the productive system, and this naturally pushes it to the left in the context of Western politics' (Berger 1987: 69). Through analyses of this sort one can make sense of the typical constituency of support for environmental movements, movements which tend to appeal to the well-educated and to those in the distributive aspects of the economy. We can, in other words, understand the customer base to which environmental SMOs appeal. But as Berger points out, the knowledge class is huge. It forms much of the basis for support for the green movement but underdetermines that support. Other people with an identical class profile to that of keen environmentalists may support different movement causes, promoting community arts of alternative therapies.

In summary, no doubt Melucci (1989: 194) is partly correct to ascribe the differences between the typically European and the typically US approaches to differences in the social and political context in the two continents. But the intuitively appealing proposition of the European tradition, that only some historic movements ought to be distinguished by the title of 'new social movements', appears to be unsustainable. No theoretical or methodological basis can be offered for identifying the exclusive claimants to this title. The environmental movement surely merits the title of a new social movement. But, while at first sight it appears to have the correct kind of transformational potential to qualify for movement status in the European sense, its transformational ability remains to be empirically demonstrated. Furthermore, competition and divisions among Greens also imply that its coherence as *a* movement must be in doubt.

THE SPECIFIC FEATURES OF THE ENVIRONMENTAL SOCIAL MOVEMENT

Accepting that the term 'social movement' should be used as an empirical generalization, one is led to ask how does environmentalism distinguish itself within this category? One characteristic will be reserved for the next section, its scientific nature. But we are still left with several sociologically important features which I will group under two headings. The first of these is the international character of the movement. Many social movements have claims to international solidarity – international socialists are one obvious and ironic example. In principle there are probably good reasons for international solidarity in all sorts of social movement; even nationalists within the former Soviet Union have reason to co-operate. Peace movements have a strong and urgent reason for international solidarity too but their unity generally foundered on the asymmetrical opportunities for peace organization during the Cold War. But environmental issues are, I suggest, unusually international for several reasons.

First, environmental threats are themselves often 'transboundary' in character. Whoever pumps them out, CFCs will affect the ozone layer. The USA or Europe cannot attend to its own interests in this matter (although this is not to say there are no disputes over the North's and the South's differing views on humankind's ecological interest needs; see pp. 165–6). The same is true for global warming, to some extent for acid rain and even for marine pollution. Rivers and water resources are typically shared between countries also, as with current concerns over access to and pollution of the Danube.

The cross-national character of the problems has meant that the respective countries' SMOs have seen opportunities for co-operation, opportunities which they have generally seized with greater alacrity than governments. Cynically expressed, governments have an interest in getting other nations' leaders to do as much about remedying a problem as possible so that they themselves have to undertake a minimum amount of work. SMOs don't obey this logic. By and large they are free to press for optimal action on environmental reform. Furthermore, they can use the green performance of the leading reforming country in any particular area of policy to berate their own governments. In the last decade, US legislation, on car exhausts and on freedom of information, has repeatedly been pointed to by European campaigners with an argument such as, 'If the US government can insist on catalytic converters and still retain a car industry, surely European governments can do the same'. This line was adopted in the infamous Greenpeace campaign which employed Ford's own slogan ('Ford gives you more') to point out that a British Ford gave you much more toxic pollution than an American Ford.

The importance of these international issues has also fed a virtuous cycle as far as environmentalists are concerned. As McCormick points out, those British groups which have the most international outlook are precisely the ones which have experienced the most rapid growth in the last decade (McCormick 1991: 151–2). Furthermore, international co-operation is also less affected by intra-movement competition than national action since international 'partners' are not often fishing in the same pools for support.

A second reason is that campaigners have often focused on aspects of the 'commons'. The two most striking examples of this are the Antarctic and the oceans. All the time that negotiations proceed over various nations' exact entitlements to Antarctica it is possible to argue that it is a common inheritance for the whole planet and ought to be specifically safeguarded (witness the fifty-year moratorium signed in 1991). Similarly, the seas outside of territorial waters have long been accepted as a form of commons; countries which dump into them, over-fish them, or contaminate them with radioactive dust or incinerator waste can be represented as fouling the joint nest. In short, it is possible to identify campaign themes which embody and symbolize the notion of common wealth. These are areas whose special status has already been internationally acknowledged in one form or another. And many nations' environmentalists are ready to subscribe to such causes. These initiative may be difficult to administer but they are a ready opportunity for international solidarity.

159

Lastly, the internationalism of the movement has been assisted because various international bodies, and in particular the EC and the UN, have fastened on to environmental issues as a way in which they can augment their influence. The environment appeals to such bodies because they can argue that it is inherently international and because it can be presented as a public interest issue. If, for example, the EC is putting forward proposals which supposedly advance the public environmental good it is hard for national politicians to oppose these without seeming to argue out of pure self-interest – seldom a rhetorician's favourite stance. Environmentalists may then find that these international bodies afford a 'softer' lobbying target than do national government. However, it should be noted that the costs and administrative demands of lobbying at this level tend to screen out smaller environmental organizations and thus to favour the larger campaign groups.

These factors explain why, in addition to the ideological reasons which Greens may have for favouring international collaboration, they have been successful in developing international solidarity. Of course, some practical features militate against this internationalism. Environmental SMOs develop a deep familiarity with their own countries' laws, politicians, civil servants and media. They are often so busy addressing these issues that the scope for internationalism is limited. Thus I am not suggesting that the green movement has transcended national barriers. But there are sociological and political reasons for believing that it stands a better chance of doing so than other putatively universal social movements; its only obvious 'rival' here is the women's movement which also appeals to supranational objectives but with fewer pragmatic supports for its universalist pretensions.

The green movement's international orientation is additionally important in relation to the treatment of environmental issues in the Third World; this is a subject I will return to later (see pp. 163–6).

For now, I shall return to one other area in which the specificity of the green movement is apparent. This is the extent to which Greens offer a critique of capitalism, an alternative value system and a view of the alternative society which they would wish to see ushered in. In other words, Greens offer a more comprehensive alternative than other social movements; this is perhaps why they are often uppermost in the minds of analysts such as Melucci and why they seem to offer, as Touraine put it, a challenge to our great cultural orientations. The peace movement does not offer such a comprehensive and radical programme; it is difficult to argue that a commitment to peace necessarily leads to an antipathy to liberal capitalism (that's not to say the argument hasn't been attempted – for instance, the notion of capitalism needing a permanent arms economy – but that it hasn't found widespread support). Certainly, the peace movement does not project an alternative blueprint for social organization. The same would be true of the youth movement. Again, the feminist movement is the closest rival to the Greens. But, to adopt the useful phrase which came into brief currency on the disintegration of the Eastern bloc, only the environmental movement offers a distinctive challenge to the idea that there is an 'end to history'. Greens have, as Dobson recently argued (1990), a coherent green political philosophy; they have distinctive views on the economy. They of all social movements have founded political parties in many countries, north and south, and experienced some electoral success. As Lowe and Rüdig note (1986: 537), 'Only the ecological movement represents a totally new political cleavage.'

However, the movement is not as coherent as this implies. As McCormick reports, the key supporters of the (UK) Green Party and of the leading environmental movements overlap only to a limited degree (McCormick 1991, 123–4), a finding reinforced in a recent survey of British Green Party members which stated that 'only a minority of Green Party members have been involved in social movement activities of some kind to a significant degree' (Rüdig *et al*. 1991: 36). Furthermore, the leading British SMOs have been careful to distance themselves from the Green Party. Greenpeace, perceived to be the most radical of the SMOs and set apart by its willingness to break the law, calls for an end to non-natural pollutants entering the air, water or soil. But it does not explicitly state that this implies a reduction in Western standards of living or a redistribution of income across the globe. Jonathon Porritt has long argued that 'we must be prepared to reduce our own standard of living' (Porritt 1988: 20), and in a variety of ways he associated this view with FoE. However, the substantive focus of much of FoE's campaigning has been on suggesting improved policies for power generation, farming or transport. This advice has principally aimed at maintaining our standard of living while lessening pollution and resource depletion. Equally, they campaigned for the removal of CFCs from aerosols – not for the total banning of aerosols. Whatever the private convictions of leading campaigners, the demands of successful campaigning dictate that environmental SMOs wear a less than deep green garb. The in-principle transformational character of the green movement is thus ironically diluted by the practical outlook of its leading SMOs.

A final aspect of this issue goes back to Berger's analysis of the knowledge class. As he himself notes, this class is in an ironical position of hostility to the productive forces which have brought it into being. Without economic surplus this class would wither away. Thus, the very class in which environmentalism finds its most active and eloquent support has material interests antagonistic to a deep greening of society. The tensions inherent in the movement are replicated in its support base.

SCIENCE IN THE GREEN MOVEMENT

The role of scientific knowledge claims and of scientific authority in the green movement is well acknowledged at a practical level. Within the movement it is demonstrated, for example, by the appointment of Jeremy Leggett as Director of Science at Greenpeace (UK) and by the amount of original research undertaken by the RSPB. It is reflected in the growth of a popular science literature among the interested public, with books such as *The Hole in the Sky* reaching a large audience. It is demonstrated by the (apparently) growing commitment of national governments to spending on environmental research. But it has not, by and large, been remarked on by sociological analysts, except at the level of generality at which theorists such as Beck write (see Yearley 1992: 113–48).

Of course, one might argue that it has not been emphasized because it is common to other social movements or pressure groups which also make extensive use of scientific arguments. Public health organizations of the nineteenth century would be a clear example. And other candidate 'new social movements' are not without their scientific aspects. Thus the idea of a nuclear winter became an important part of the peace campaigners' case, while a robust attitude to biologists' arguments was central to the development of feminism. However, I suggest that science is of special importance to the green case. For one thing, many of the objects of environmental concern are only knowable through science. Without a scientific world-view we would know nothing of the ozone layer and would certainly be unable to measure its diminution; the same is true of the greenhouse effect. But science is important in other ways too. For example, scientific reasoning gives us the notion of a sustainable population size for endangered species; natural history instructs us in the needs of animals and plants (what or whom they need to eat, where they like to nest, the hazards to which they are susceptible); scientific reasoning even allows us to seek to specify the risks from toxic pollutants.

This is not to say that science has been a dependable friend to the environmental movement. In some ways the contrary is true. As is well known, the demand for scientific 'proof' has been used to justify official inactivity towards environmental problems (over the abatement of acid air pollution for example); the authorities have argued that they should wait until the evidence of damage is conclusive before making major policy changes. The scientific community, particularly in the nuclear industry, has often appeared committed to activities which heighten environmental hazards. Threats to the environment have often stemmed from technological progress; after all it was scientific advice which led to the addition of lead to petrol and to the widespread use of CFCs.

Both sides in environmental debates have typically tried to enlist the support of science so it makes no sense to try to say which 'side' science is on. However, one can note that the central role of scientific argumentation in the green case has had an important impact on the sociology of the movement. First, the importance of scientific expertise has added to the distance between the professional cadres in the SMOs and the casual movement supporter. While campaign professionals learn more and more about acid pollution or radioisotopes, 'ordinary' members are encouraged to deal chiefly in terms of general principles such as the 'polluter pays' or 'precautionary' principles. One should not try to make too much of this observation. Clearly full-time peace campaigners became experts on multiple independently targeted re-entry vehicles (MIRVs) and 'Star Wars' technologies, while many 'ordinary' Greens are very knowledgeable about ecological issues. None the less, the scope for, and importance of, scientific expertise in Greens' arguments are particularly great.

Second, the centrality of scientific evidence holds out the beguiling possibility of winning one's case simply through argument. When Margaret Thatcher seemed to have been persuaded by the growing evidence on global warming (and commentators noted that *after all* she had trained as a scientist), it appeared that the demonstrable correctness of the case might set environmentalism apart. This notion appealed in particular to the more 'establishment' environmental groups, such as the Royal Society for Nature Conservation (RSNC) which, throughout its eighty-year history, has had a predominantly scientific ethos. The supposedly universalistic characteristics of scientific argument thus lent credibility to one style of campaigning and has left its mark on the composition of the overall green lobby.

ENVIRONMENTALISM AND SOCIAL MOVEMENTS IN THE UNDERDEVELOPED WORLD

Earlier (pp. 158–60) I reviewed reasons for thinking that the green movement would be unusually successful as an international movement. However the majority of considerations I reviewed there concerned only the countries of the industrialized West. Some of those arguments can be readily extended to include the Third World – global warming, ozone depletion and dwindling biodiversity, for example, are supposedly truly universal threats and ought to be a stimulus to action everywhere. But how sociologically accurate is the idea that the 'universality' of ecological problems can lead to a unity of interests between Greens in the First and Third Worlds?

Writers on ecological issues in developing countries have long noted a serious problem: countries which are eager for development are often faced with an apparently stark choice between industrialization and environmental protection. Such countries have often succeeded in attracting investment precisely because they have a greater toleration of pollution or of environmental hazards. Some critics have argued that an over-fastidious concern with the environment will condemn the Third World to continued poverty. This conviction, particularly in the Third World itself, appears to be changing and it will be helpful to review the reasons why this may be.

In some cases ecological destruction is simply so extensive, due to loss of soil, water contamination or desertification, that it obviously denies the opportunity for development. But elsewhere the connection between environmental change and people's beliefs about development options has been more complex and mediated. One experience which encouraged politicization around environmental issues was the common acknowledgement, in the late 1980s, that several West African countries were being used as dump sites for the West's unwanted and dangerous waste. It became clear that these countries were unlikely to suffer the environmental costs of processes from which they had not even had the economic benefits. Through the Organization of African Unity these states were able to act in concert to limit this trade and to monitor each other's performance. Still, the practice is continuing in other parts of the world, with Latin American countries currently increasing their waste disposal facilities – often with the assistance of the World Bank and in the face of local public opposition.

A second, and related, stimulus to environmental politicization was the strategic choice by Western firms to locate polluting and hazardous plant in investment-hungry Third-World countries. The tragic accident in Bhopal came to symbolize the exploitative and seemingly heedless attitudes which lay behind such investments. And it was not only the actions of private firms which displayed this connection. Officially aided development schemes and even aid programmes had negative and sometimes disastrous environmental implications. Such incidents again changed the perceived balance in the trade-off between industrialization (or development) and environmental protection.

A third way in which the ecological message has come to be accepted is because of ironical feedbacks in the economies of Third World countries. For example, Caribbean islands are eager for development and are consequently keen to bring in industries. But they know that the dirty industries which it may be easy to attract are going to conflict with their potential for tourist development. Similarly, in the Irish Republic the Industrial Development Authority (IDA) has found that certain developments have been opposed by local groups who have well-worked-out arguments demonstrating that earnings from tourism, or the fishing revenue or even traditional farming practices will be affected by the new industry (see, for example, Allen and Jones 1990: 9–10). And since these things are enduring sources of economic prosperity, while factories sometimes stay only a few years, it can be said that they offset the apparent advantages of industrial development.

These mechanisms have encouraged belief that ecological concern is central to the politics and practice of development, and community-based environmental movements are now well established in many Third World countries (see Redclift 1989: 159–70, and Pearce 1991: 157–81). However, we still need to ask whether the potentially universalistic character of environmental problems has been accompanied by actual solidarity. One might suppose, for example, that such solidarity might be expressed through responses to transboundary ecological problems. But in many cases these have been perceived as attempts by First World Governments to solve their own problems at the expense of the Third World's developmental potential. Thus, proposals to limit carbon dioxide emissions from the Third World which are not matched by drastic reductions in the First World's output are seen as merely hypocritical. Moreover, as McCully has recently argued in *The Ecologist*:

> Demands for huge amounts of climate aid for the Third World give First World governments an excuse to do nothing except point out to their electorates the huge costs of dealing with the problem and enable them to shift the blame for global warming from the historic and present high emissions from industrial countries onto the projected future emissions from the Third World. It also gives Third World governments the excuse of not doing anything because they can claim it is too expensive and the First World will not stump up the money.
>
> (McCully 1991: 251)

Through suggested official initiatives such as climate aid the movement is being offered measures of an apparently universalistic nature which may not actually operate in anything like the common interest.

The World Bank's Global Environment Facility (GEF) has been subject to similar criticism (see Ticknell and Hildyard 1992). Two sorts of argument have been made. First, the GEF can go to redressing the environmental damage caused by the sort of development projects the Bank might well have funded before ecological issues came to the fore. For instance, the Facility could be used to lessen the environmental damage associated with dam construction. Rather than preventing ecological damage, it seeks to compensate for it. Implicitly, the critics argue, this means that the GEF will allow environmentally damaging projects to go ahead which, in the absence of the Facility, might have been opposed. Second, and more subtly, it is argued that the GEF only directs funds to ecological problems which are construed as 'global'. But the globality of environmental problems is not an inherent or obvious quality; 'global' status is itself constructed. And the Bank's Southern critics argue that it has been constructed to coincide with those environmental problems seen as most pressing in the North, and not those – such as poverty – which the South might try to construe as global.

The problems of universality are not confined to official agencies. The apparent urgency of the world's ecological problems and the seemingly universal benefits of suggested solutions can also lead Western environmental SMOs into actions perceived as supercilious in the Third World. In some respects there is an obvious parallel here with current debates within feminism, where the right of Western feminists to speak for womankind is questioned on the grounds of race, ethnicity and financial privilege. Some contemporary feminists look to the 'celebration of difference' said to be encouraged by post-modernism for a solution to this difficulty. But the analogous move is not open to environmentalists who generally find it hard to 'celebrate' differences in views about the ozone layer. It is difficult for First World campaigning organizations which have the ear of the media and considerable expertise to find ways of facilitating participation by Third World groups. The dynamics of this interaction have become clearest over attitudes to tropical rain forests where the scientific and economic issues leave much more room for local negotiation than is the case with, say, ozone-depleting chemicals.

Environmental SMOs in Third World countries have a different life history from those in the West. They may owe their development to the influence of external organizations such as aid NGOs or to overseas environmental NGOs. Their relations to local media and to political actors will be different. They may face environmental challenges of a different sort from those common in Western Europe where dirty foreign industries, agribusiness and plantation agriculture are much less common and where they are not usually the prime targets of environmental campaigners. Only an open and empirical definition of 'social movement' will allow them to be systematically compared with green SMOs in the North.

CONCLUDING REMARKS

In this overview I have sought to explore the contribution of the sociology of social movements to our understanding of environ-mentalism and social change. I argued that only an essentially descriptive definition of 'social movements' is acceptable; however, leading features of the environmental movement are well captured by this definition. Although there are many similarities between environmentalism and other movements I identified three ways in which the green movement stands out: its intimate relationship to science, its practical claims to international solidarity, and its ability to offer a concerted critique of, and alternative to, capitalist indus-trialism. According to Lowe and Rüdig (1986: 537) green politics has a peculiar inclusiveness and exclusiveness. It can accommodate other radical movements while resisting assimilation into their ideological agendas. I suggest that the three factors I have highlighted account for this property. It should be noted though that the claims to univer-salism offered by the green movement and its critique of Western society – however coherent in principle – are in practice limited. Leading environmentalists compete too much and distance themselves from deep-green politics too much for the movement to have the practical coherence which inclusiveness and exclusiveness seem to imply; in this regard I suggest Lowe and Rüdig's claim is exaggerated. Last, there are special grounds for doubting whether the movement's avowed universalism allows First World environmentalists to represent the Third World in an unproblematic way. The special conditions shaping social movement formation in the Third World are still little researched, as are relationships between Third World environ-mentalists and aid NGOs and environmental SMOs in the industrialized world.

ACKNOWLEDGEMENTS

I am very grateful for the helpful but critical comments on this chapter offered by participants in an ESRC-sponsored Global Environmental Change Workshop at the University of Kent at Canterbury, by participants at a seminar at the School of Social Sciences at the University of Bath and by my colleague Arthur McCullough. I fear I may still have paid too little attention to their many comments.

REFERENCES

Allen, R. (1992) *Waste Not, Want Not*, London. Earthscan.
—— and Jones, T. (1990) *Guests of the Nation*, London: Earthscan.
Beck, U. (1992) *Risk Society: Towards a New Modernity*, London: Sage.
Berger, P. (1987) *The Capitalist Revolution*, Aldershot: Wildwood House.
Dobson, A. (1990) *Green Political Thought*, London: Unwin Hyman.

Eyerman, R. and Jamison, A. (1991) *Social Movements*, Cambridge: Polity.

Giddens, A. (1990) *The Consequences of Modernity*, Cambridge: Polity.

Jamison, A., Eyerman, R. and Cramer, J. (1990) *The Making of the New Environmental Consciousness*, Edinburgh University Press.

Lowe, P.D. and Rüdig, W. (1986) 'Review article: political ecology and the social sciences – the state of the art', *British Journal of Political Science*, vol. 16, 513–50.

McCormick, J. (1991) *British Politics and the Environment*, London: Earthscan.

McCully, P. (1991) 'The case against climate aid', *The Ecologist* 21(6), 244–51.

Melucci, A. (1989) *Nomads of the Present*, London: Hutchinson.

Newby, H. (1991) 'One world, two cultures: sociology and the environment', BSA Bulletin *Network*, vol. 50 (May): 1–8.

Pearce, F. (1991) *Green Warriors: The People and the Politics Behind the Environmental Revolution*, London: Bodley Head.

Porritt, J. (1988) 'Greens and growth', *UK CEED Bulletin*, vol. 19. 22–3.

Redclift, M. (1989) *Sustainable Development*, London: Routledge.

Rüdig, W., Bennie, L.G. and Franklin, M.N. (1991) *Green Party Members: A Profile*, Glasgow: Delta Publications.

Tickell, O. and Hildyard, N. (1992) 'Green dollars, green menace', *The Ecologist* 25(3), 82–3.

Touraine, A. (1981) *The Voice and the Eye*, Cambridge: Cambridge University Press.

——, Hegedus, Z., Dubet, F. and Wieviorka, M. (1983) *Anti-Nuclear Protest, the Opposition to Nuclear Energy in France* (trans. P. Fawcett), Cambridge: Cambridge University Press.

Yearly, S. (1992) *The Green Case*, London: Routledge.

Zald, M.N. and McCarthy, J.D. (1987) 'Social movement industries: competition and conflict among SMOs' pp. 161–80 in M.N. Zald and J.D. McCarthy (eds), *Social Movements in an Organizational Society*, New Brunswick: Transaction Books.

8

SCIENTIFIC KNOWLEDGE AND THE GLOBAL ENVIRONMENT

Brian Wynne

INTRODUCTION

In sociological company it does not need to be argued that 'the global environment' is a social and cultural construct, into whose multivalent articulation are poured many complex and conflicting anxieties and commitments. The discipline of scientific knowledge is seen as the one superordinate discourse which can lend coherence to this incipient anarchism, to identify and describe the real natural problems, account for the underlying processes, and to define reliable and realistic options for societal response. Thus the social authority of science becomes a central issue, and ever more sharply so that the environmental and geopolitical arena over which it is supposed to reign expands to literally global proportions.

Two features of this developing global environment enterprise and its associated problems of global authority beg attention here. The first is the concentration on developing a 'sound scientific basis' (UK White Paper 1990) for internationally agreed policies, notably via the working groups of the Intergovernmental Panel on Climate Change (IPCC). The second is the explicit integration of social science into the framework of science and policy. In principle both of these processes can be taken for granted. However, the particular forms which these principles are being given as they are translated into practice are also taken for granted. Sociology of scientific knowledge has a role to play in exposing some neglected issues in both these aspects.

The way in which social science has been brought into the global environmental change programmes of virtually all the significant national and international research and policy bodies (from the UN to the US National Science Foundation and the Human Dimensions of Global Change programme of the International Geosphere Biosphere Programme, with the exception of the UK Economic and Social Research Council (ESRC), but not of the UK Inter-Agency Committee

169

for Global Environmental Change (IAC)) has been as a subordinate to the precommitments and agenda of the natural sciences. It either provides information to the natural sciences on human activities which perturb the natural processes, or takes the natural science predictions as given and then works out the social and economic consequences.

A further role entirely consistent with this subordination to the basic agenda and epistemology of the natural sciences arises from the authority problem. Here social science is supposed to offer ways of educating global publics into better understanding and appreciation of the 'real' hazards (for example, ASCEND 1992), the questionable assumption being that lack of public uptake of scientific knowledge and prescriptions is based only upon ignorance or misunderstanding, not upon any more fundamental problems of cultural identification or alienation. Some of these deeper difficulties may be associated with the particular 'cultural' properties of science which are exposed by a sociological examination of what is meant by, and automatically assumed to be, 'sound science' for global environmental policies.

Although this subordinate relationship to natural science has been questioned (and rightly so), the reasons for doing so have not always been so clear. If as some including myself have argued, social science ought to be involved more upstream in the currently 'private' processes of constructing scientific knowledge or technological artefacts, how far does this sociological remit properly go, and what are the implications? If environmental scientific knowledge is sociologically deconstructed, what does that deconstruction itself 'reveal'? And what is the relevance of the criticism sometimes aired that sociological treatment of the current scientific 'consensus' that we have a greenhouse warming problem inevitably plays into the hands of the international fossil fuels lobbies and their political friends who are doing their very best to demolish that consensus?

In this chapter I will suggest that there are confusions on these questions, and that these relate to the long-standing differences between interests-based and more culturally rooted social constructionist perspectives on scientific knowledge.

The issues for sociology of science in relation to the global environment can be partly summed up in the question: What social or cultural factors do particular scientific discourses about the global environment tacitly reflect as if natural, and thus exempt (whether deliberately or not), from wider debate, negotiation and responsibility? In his book, *Strange Weather*, Andrew Ross has suggested that the structure of US weather forecasting is 'shaped by a social and political mapping of the world as much as it is determined by the atmospheric map of shifting fronts and air-masses' (Ross 1991: 240). What kind of social and political maps are implicitly represented in global

SCIENTIFIC KNOWLEDGE AND THE GLOBAL ENVIRONMENT

environmental science? Posed from a social constructivist perspective this question also includes a challenge to social science, because the boundaries of the 'natural' and the social or cultural are at issue, and much of social science as currently engaged in the global environment field is obscuring the full scope of the social and cultural issues by default.

REDUCTIONISM AND THE
GLOBAL ENVIRONMENT

It is ironic to note that as the geopolitical reach of environmental science has become more and more expansive, its intellectual temper has become more reductionist. Whatever the justifications may be for globalizing the instrumentalist and standardizing culture of science, the problems of diverse local cultural identification and authority for its pronouncements and associated policies only escalate even further as a result. Conventional approaches such as those emanating from scientific institutions see the problems here to be about public ignorance, and the need to develop more rational attitudes (for example, ASCEND 1992). However sociological approaches recognize that scientific knowledge reflects social and cultural factors which render it parochial in important respects; just how parochial in the global political and cultural milieux, remains to be charted.

In recent years the most prominent focus of 'official' scientific attention to the global environment has been the IPCC and its scientific working groups attempting to give objective projections of world climate futures and their dependence on anthropogenic carbon and other emissions. Newby, amongst others, has noted the apparent regress from the 1987 UN Brundtland Commission's broad focus on global environmental futures, to that of the more recent IPCC. Whereas Brundtland articulated a basic political, moral and social framework from which to define policies for environmentally sustainable global development, including scientific R&D and technology policies, IPCC began from a scientific origin – defining and managing a sustainable climate – from which should be derived the necessary social, economic and other policies for survival.

Newby's observations resonate with wider critiques of both Brundtland and IPCC, that they still, albeit in different ways, reflect rich world agendas and interests, and limit via 'natural' expert authority the uncertain, but certainly uncomfortable, extent to which the industrialized rich world needs critically to examine and reconstruct its own deep commitments in order adequately to address the contemporary crisis which is called global environmental change. Various authors have commented on the evident point that the IPCC

171

science-led process has effectively reduced the global change problematique from the wider political economy issues surrounding North–South inequities, the debt burden and the 'environmental poverty trap' in which poor countries are in too desperate an economic state to be able to forgo immediate resource exploitation whatever the long-term implications for sustainability. Whereas the IPCC's discourse is of a global *environmental* crisis, the critical voices assert a global political, economic, moral and cultural crisis. The question of interest here is whether, and how, 'scientific' discourses such as that around the IPCC prevent us from recognizing and responding to the awful challenges of the larger version of the global crisis.

The political perspective on global environmental change leaves the science *as knowledge* untouched by critical analysis. That it is alleged to be handmaiden to political attempts by the rich countries to impose a policy consensus on the developing world implies nothing about the kind of scientific knowledge being developed and deployed. That IPCC equates global environmental change with climate and greenhouse warming, and mainly with carbon emissions, is already a form of reductionism with political implications. There is also some analysis (for example, Lunde 1991) of the management of scientific consensus in IPCC in order to try to achieve greater political authority. However, the more subtle and difficult question is whether the deeper epistemological commitments on which the scientific knowledge is built serve to constrain the vision of what is at stake, socially and culturally.

One useful approach to this question is via historical reflection. In the present scientific enterprise, 'sound science' is taken to mean huge supercomputer models which are only available at six research centres in the world. These mathematical titans involve physical meteor-ological parameters, equations and sub-models. They are the direct descendants of weather forecasting models which extended their scope from a few days to the outer envelope of viable predictive control of the relevant variables, thought to be about twelve days, and are now extended to try to give credible predictions thirty to forty years hence, based upon given input assumptions about atmospheric carbon emissions.

As many have noted, despite their colossal size and complexity these models leave out several important factors such as cloud behaviour in relation to global warming, and biological processes such as marine algal fixing of atmospheric carbon, natural methane production and release as temperature changes. There is no reason to suppose that these processes are any less basic to global climate change mechanisms than the processes which have been incorporated, it is just that they have not been treated as central in weather forecasting and so have not been previously subjected to intensive data gathering – at least not by recognized scientific research specialities.

The physical-mathematical models are being validated, which also means examining the significance of these omissions, by comparing their outputs of calculated global temperatures from retrospective runs of the models with past data on global temperatures.

These reconstructions of past temperature and climate states, stretching back centuries, have been performed by a range of different research disciplines, from geography and social history to geochemistry and palaeogeology. The kind of evidence adduced ranges from parish records of crops and harvests, to radiocarbon dating of ice-cores to establish dates of inferred measures of temperature and carbon dioxide at different sites. This kind of science is manifestly indirect in its access to its object, namely data on temperature and CO_2 levels and distributions. It relies openly upon indirect and surrogate variables, and composite variables, where control for other factors, known or unknown, is overtly more difficult.

Although it is now dependent upon this non-reductionist scientific knowledge for validating its own mathematical models, the meteorologists and physicists who dominate the science-policy climate-change domain have long regarded it as unsound. In the traditional cultural hierarchy, good science is taken to equate with that which allows prediction and control with analytical atomization, high precision and single-variable measurement and manipulation where possible. The Meterological Office, whose Director headed the scientific working group 1 of the IPCC, has been the guardian of this concept of good science in the climate-change field. It is overwhelmingly staffed by mathematicians and physicists, and has in turn dominated UK research on climate. When the geographer Hubert Lamb, who had been on a special fellowship at the Met. Office, left in the early 1970s to establish the Climate Research Unit (CRU) at The University of East Anglia, with the explicit interest in long-term climate changes and reconstruction of past climate states, financial support was not provided by the UK research system. The Met. Office opposition to Lamb's kind of science found resonance in the reductionist culture of 'good science' generally in the UK, and the CRU programme was only rescued thanks to private support from Shell International and The Nuffield Foundation.

There is a particular irony in this situation which is worth emphasis. At this time, in the 1970s, the Met. Office scientists did not believe that long-term climate change was an issue. Thus whilst Lamb and colleagues gathered their 'unsound' scientific evidence for reconstructing long-term climate changes, Mason and colleagues from the Met. Office were publishing rebuttals (see, for example, Mason 1976) to the effect that the global climate system contained strong equilibrating factors which made it resilient against perturbations; fluctuations occurred, but only around a stable mean climate condition. Climate change was thus a grossly exaggerated issue, according to the meterologists, and the broader climate researchers such as Lamb were responsible for the exaggeration.

Ironically, therefore, the discipline – and indeed the institution – which now dominates the scientific and policy consensus-making on global climate change was then pouring scorn on other 'unsound' sciences which were advancing the idea that climate change was a problem, even if the causes and directions of change were still obscure. There were even hints that these climate-change scenarios and their associated disciplines were part of a pernicious antiscience movement. Certainly they enjoyed widespread public uptake, as in Nigel Calder's 1970 BBC TV programme, *The Weather Machine*; and the diatribes against the emergent environmental movement as 'antiscience' (for example, Maddox 1972) included reference to the environmentalists' belief in climate change and the anthropogenic greenhouse effect.

Thus the less reductionist perspective of the other sciences on global warming was initially denigrated and denied as 'second-rate science' by the scientific establishment, only to be taken over and transformed when the notion of climate change gained wider credibility, into the reductionist idiom of the powerful supercomputer models of the IPCC.

The dependency of these models upon longer-term historical data to validate them does not alter this state of affairs, indeed reports from IPCC negotiations indicate that these surrounding disciplines are organized as servants to fuel the megamodels. So too it appears are the scientists from institutions in the developing world – and even in the developed countries – who do not have direct access to the few supercomputer models (Liverman 1991). There are important implications for the credibility and purchase of policy proposals seen to be emanating from this narrowly exclusive cadre.

Resistance to the kind of climate research at CRU on the part of the UK research policy establishment was based upon the physics-based notion that sound science equals reductionist, high control, high precision science; thus the science of CRU and Lamb was unsound, and unworthy of support – 'second-rate science' (Beverton 1992).

This same reductionist cultural notion of sound science in environmental policy is institutionalized in UK research policy. For example, a study of UK environmental research (Wynne 1991) found not only that the majority of funding went to physics-based research projects, but that even within biological research as a whole a majority of funding was in laboratory-based molecular biology and genetic manipulation – the biological counterparts of physics. Some researchers

174

recognized that this artificial, high-control research had exhausted its possibilities because of the lack of research actually testing behaviour in (less reductionist) field conditions.

Thus the cultural syndrome identified in the competition between physical meteorology and other less-reductionist disciplines over how to define and develop climate change knowledge was part of a wider pattern (also affecting social science) that begs further research. The recognized disciplinary paradigm conflicts between ecology and physics in science-policy disputes are part of this deeper picture yet to be properly filled out. Robbins and Johnson's (1976) case-study of toxicology versus geochemistry in the environmental led controversy in the early 1970s is a similar example.

For the climate research case these considerations raise questions already aired in relation to other science policy areas, such as environmental and safety regulation, as to how criteria of good science for policy ought to be defined (Jasanoff 1990), and whether the unreflective adoption of the dominant criteria in scientific institutions is an adequate basis for public debate. In other words, do the competing styles of science reflect different epistemological, institutional and cultural correlates? And do these contain different implicit prescribed boundaries, not only to the science-policy interface but also to the extent of problematization of the human subject or to cultural identities embedded within the processes of global change?

WHY SOCIOLOGY OF SCIENTIFIC KNOWLEDGE? RETRIEVING INDETERMINACY

Although the reflex reaction is still to understate and, where possible, conceal scientific uncertainties in public policy issues, it is now commonplace to find the inevitable limitations of scientific knowledge recognized as a fact of life which policy-makers and publics should learn to accept (Ravetz 1990, Smithson 1989, Jasanoff 1990, Nelkin 1979). Thus scientific uncertainty is widely discussed as the cross which policy-makers have to bear, and the main obstacle to better and more consensual or authoritative policies. Yet much of this debate still assumes that if only scientific knowledge could develop enough to reduce the technical uncertainty, then basic social consensus would follow, assuming that people could be educated into the truth as revealed by science.

There are two main sociological strands of criticism of this dominant conventional perspective. The interests-oriented strand would note that even within the constraints of an accepted natural knowledge consensus, legitimate social interests – and hence favoured policies – can be in conflict. A perspective from the sociology of knowledge would go further, to argue that dominant interests control expertise and hence shape the available knowledge to reinforce their interests.

A more radical strand would suggest that beneath the level of conflicting explicit preferences or interests lies a deeper sense in which scientific knowledge tacitly reflects and reproduces normative models of social relations, cultural and moral identities, as if these are natural. Thus, for example, the level of intellectual aggregation of environmental data and variables such as radiocaesium in the environment, when used to establish and justify restrictions on farmers operating in that environment, is effectively prescribing that degree of social or administrative standardization of the farmers.

In other words, at a deeper level than explicit interests the form in which scientific knowledge is practically articulated prescribes important aspects of their social relations and identities. In research on the interactions of scientists and farmers after Chernobyl, this point came out as the farmers' detailed and differentiated local knowledge of the environment and what it meant for optimal farming methods, even in the same valley, were denied by scientific knowledge whose 'natural' form aggregated and deleted them into single, uniform data categories combining and homogenizing several different valleys and many farmers. As one farmer caught by the Chernobyl restrictions lamented in this respect: 'this is what they can't understand; they think a farm is a farm and a ewe is a ewe. They think we just stamp them off a production line or something.'

This brief glimpse indicates that the scientific knowledge is not naturally determined; it could have been organized differently and still have respected the evidence from nature. Yet social commitments to such organizing epistemic principles as the levels of aggregation of entities into uniform conceptual classes and categories are so deeply enculturated into the scientific canons of given specialities or fields that they are mistaken as being completely determined by nature.

The indeterminacy which this kind of example exposes is fundamentally different from the uncertainty normally bandied about in discussions of science and policy. In trying to make sense of scientific consensus building in the context of the IPCC and global environment politics for example, Lunde (1991) tries to distinguish between 'standard' epistemic factors in consensus building, such as empirical observation, measurement and monitoring, and validation of models; non-epistemic factors, such as political interests and lobbying; and 'non-standard' epistemic factors, such as saliency and the perceived social role of knowledge. The problem with these distinctions is that they are exposed as vacuous as soon as they are applied to concrete instances. Sociology of scientific knowledge has shown repeatedly, and often in great detail, how a sacred canon of scientific method such as the replication of empirical observations – another 'standard epistemic factor' – is a fundamentally underdetermined normative principle 'controlling' scientific knowledge building. The same is true of inference rules and logical commitments which define entities as belonging to the same class or different collective categories depending upon which properties are taken as salient. The actual meaning of these 'natural' terms and rules have to be negotiated as research goes along. This is a fundamentally more open-ended process of knowledge construction than is recognized in conventional perspectives, which treat scientific knowledge as fully determined by nature alone, and which correspondingly treat scientific uncertainty as a kind of temporary pathology awaiting more rigour or precision which will supposedly reveal the 'true' determinism underlying things.

Radiocaesium behaviour in the environment

An example of the deep indeterminacies concealed in scientific knowledge can be drawn again from the post-Chernobyl emergency in 1986. This involved scientific knowledge of the environmental behaviour of radiocaesium, which is still important in trying to manage the fall-out from Chernobyl in hill sheep farming areas like the Lake District. It is now recognized that the scientific understanding on which policy was based at the outset of the crisis in summer 1986 was mistaken; the prevailing belief was that the radiocaesium deposited from rainfall on vegetation would be washed off by further rain into the soil and, chemically immobilized, would thus no longer be available for uptake into the sheep. The sheep would thus suffer only a 'one-pass' exposure and their contamination would therefore decrease rapidly.

However, this confident belief was based upon behaviour in clay soils where absorption onto aluminosilicate clay molecules does indeed take place. No allowance was made for the conditionality of this scientific belief upon the particular soil-type, so that the very different organic, acid peaty soil conditions of the fells was ignored in the implicit view that the prevailing scientific model was universal. On this basis scientists advised that the restrictions would not be necessary at all, but if they were established against this reassuring prediction, they would only be needed for three weeks or so. Nearly 150 Lake District farms were still restricted by radiocaesium levels from free sale of lambs in 1992, six and a half years later.

Following the unexpectedly long duration of the contamination – much longer than was originally predicted by scientists – and the costly and disruptive restrictions upon farmers, allegations were made by environmental groups and some farmers that the scientists had in fact

177

known all along that the radiocaesium would remain mobile in the upland soils. They were thus alleged to have known that the restrictions would have to last much longer than three weeks, but to have entered into a conspiracy with government officials to conceal this from the farmers. These allegations were based upon the claim that the scientists had done research on radiocaesium behaviour in different soils, and had found the different behaviour as long ago as 1964. A paper published in *Nature* in 1964 was referred to in evidence (see Gale *et al.* 1964). This paper does indeed report the results of mobility tests with radiocaesium in six different soil types, including the two in question, but its reported results are ambiguous in a way which it is instructive to examine more carefully. The tests performed were of the physical mobility of radiocaesium by downward migration of a standard surface deposition after yearly time intervals of up to five years. Several samples of each soil type were set up, as is normal for this kind of experiment. The range of depth measurements at each time interval was wider for the acid peaty soils samples than for the clay (and the others), but the mean depths were the same. Thus if one took a measure of variance the soils behaved differently, but if one took mean depths the soils apparently allowed radiocaesium the same mobility. The latter would be grounds for saying that prevailing understanding was that all soil types behaved the same with respect to caesium mobility, hence in 1986 it was reasonable to extrapolate from clay soils to upland peat soils in making the post-Chernobyl predictions and policies.

Whatever the pros and cons of this political argument however, it is beside the present point. When one examines the 1964 *Nature* paper it is evident that the whole context of the reported research at that time was the fall-out from atmospheric weapons testing. The assumption then was that the critical exposure pathway to humans was not root uptake from the soils and into vegetation, thence into sheep meat from grazing, but that it was an external physical gamma radiation dose to a person on the surface direct from the radiocaesium in the ground, with shielding (reduction of the exposure) dependent upon the depth of the radiocaesium in the soil. This assumed exposure model, a social 'choice', based only on physical processes and parameters, was what determined the 'natural' scientific interest in physical depth profiles.

Yet in the post-Chernobyl sheep crisis a completely different exposure pathway came into dramatic focus, in which the key parameter was not physical migration by erosion and leaching downward in the soil, but chemical mobility and chemical availability of the radiocaesium for uptake by the roots of the ambient vegetation. Thus the scientific knowledge created by the 1964 paper, that the soils behaved in the same way, depended upon observation of *physical* parameters, a commitment dictated by the exogenous model of the critical exposure pathway assumed to exist (and probably also not unconnected with the fact that research and scientific advice was dominated by physicists, a syndrome still recognizable, though less acutely so, today). An exposure pathway of the kind identified after Chernobyl would have encouraged interest in *chemical* mobility, and would have then looked at chemical parameters instead. In this case the soils would have been different.

Thus, depending on different exogenous and contingent *social* models of exposure route, the scientific knowledge of the 'natural' properties of soils and radiocaesium could change diametrically from being 'the same' to being 'different'. These contradictory knowledges were not identified or confronted, and neither were the different conditions of validity of each stance clarified; the ambiguity and open-endedness was just left there in the wake of 'scientific progress', unresolved. The external, essentially social issue of which exposure pathway should structure research was subsumed into the 'natural' assumptions of the scientific domain, as if it were merely a scientific question. It was 'answered' by default, without anyone realizing that an arena of *social* responsibility and choice existed *prior* to and within the science. Lack of reflexive capacity on this dimension led science into the kind of trouble it encountered when its confident public predictions went wrong after Chernobyl.

This is not meant as an exercise in bashing science with the wisdom of hindsight. The aim is to show the intrinsically open-ended character of 'natural' choices and intellectual commitments embedded in, and shaping, the 'objective' knowledge outputs of science. It shows not only that what scientists knew at any given time is fundamentally problematic but, more important in this context, that even a taken-for-granted procedural principle within the research practice, such as the 'natural' significance of physical depth parameters, is a function of an externally defined commitment to a particular social scenario of exposure. Whether this is invisibly subsumed by default in the realm of science, decided upon by some science policy experts, or more widely debated and defined in society or political institutions, is itself a contingent matter which affects (and is affected by) where the boundary of science and policy is thought to be.

The precautionary principle

The development and interpretation of the precautionary principle in environmental science and policy also poses similar questions about dominant views of the relationship of scientific knowledge to policy values. The precautionary principle has been advanced as part of the general strategy of prevention, to avoid the problems associated with placing the burden of proof on the environment, and in that waiting for the evidence of harm, when remedial action is very expensive if not actually impossible, committing to policy intervention.

The standard way of defining the precautionary principle is to say that, because evidence of harm is uncertain but the error costs are very large, in some circumstances it is justifiable to intervene to protect the environment before all the evidence for a more confirmed cause-effect relationship can be gathered if it is 'reasonably anticipated' that an environmental discharge will be irreversibly harmful.

The UK government's stated acceptance of precaution closely reflects this perspective. There are two points to note. First, it retains a deterministic version of scientific uncertainty – namely, that this is a temporary matter of imprecision which will be eradicated when enough research has been devoted to the questions. Second, as a corollary, it assumes that the precautionary principle simply means moving the regulatory threshhold for intervention up the body of existing scientific knowledge – away from nature, as it were, and towards the potential polluter. *However, in the process of shifting the external policy values applying to the knowledge, that knowledge itself does not change its supposedly naturally determined internal shape.*

However, the example drawn from the radiocaesium soil-research knowledge indicated that 'when scientific knowledge knows what?' is more fundamentally open-ended, soft and thus more deeply problematic than this model recognizes, even when it expressly adopts a more precautionary standard. As we shift the normative rule through the body of scientific knowledge in this way, that body of knowledge itself may (or may need to) change as a function of the change in social or cultural values. The external normative 'choices' also tacitly influence the 'internal' choices of salient parameters, inference options, sameness and difference relations in theoretical models, and what is defined scientifically as problematic or not. On this point we can relate the radiocaesium-soils example to the production of competing kinds of environmental scientific knowledge – on the one hand, conventional research recognized under environmental assimilative capacity approaches to marine pollution regulation and, on the other, that underpinning the precautionary principle.

Dethlevsen (1988: 281) has alluded to deeper cultural differences pervading the two competing scientific approaches in his comment that 'workers who cannot see the correlation between pollution and diseases in their studies are with the exception of Moller from Germany, living on the other side of the North Sea'. But the point is that the scientists involved are not merely looking at the same body of data with different evaluative spectacles as it were, and then advising policy-makers of their policy-related judgements. Their epistemic, theoretical and methodological commitments (which may or may not include an explicit view about precaution) build up different bodies of 'natural' data or facts, impregnated with incompatible 'natural' logics, well before the policy actors come even to see, let alone exercise, normative choices about how strictly to regulate polluting activities. Thus some normative choices will have already been obscured.

Normative responsibilities and commitments are concealed in the 'natural' discourse of the science, indicating the fundamentally social, negotiable definition of the boundary between science and policy (Jasanoff 1990). The full range of moral and social issues at stake is *not* adequately described by treating the 'factual' scientific realm as if it is a separate objective 'black box' from the normative. It already reflects and expresses tacit normative boundaries and constraints.

The traditional and still dominant approach to marine pollution is a 'high-science' of the sea, akin to the reductionist idiom dominating climate research. It defines specific end-points such as fish toxicity, and focuses attention on single-variable cause–effect questions, defining environmental assimilative capacity as the maximum pollutant load the environment can carry before observable harm (defined by the choice, end-point variable) is done.

The precautionary scientific idiom is much more ready than assimilative capacity to accept:

1 that the composite variables, such as 'immunocompetence' 'disease' and 'stress', are legitimate components of scientific reasoning. Sindermann (1984) identified eighteen different factors, some natural, some anthropogenic, which might singly or combined result in stress; and 'disease therefore has to be understood to be an unspecific response towards all kinds of stress' (Dethlevsen 1988: 276). This idiom thus uses composite variables flexibly, recognizing the possible constituent factors but not discounting the larger picture just because the precise constituent variables in a composite such as 'stress' may not be defined);

2 the scientific legitimacy of indirect and multiple cause–effect inferences. For example, Dethlevsen (1988) reports a study of the possible correlation between diseases and marine contamination. Although no *direct* correlation was found, bacterial levels in the blood of eels from a contaminated area of the North Sea averaged 80 per cent compared to 4 per cent in eels from a relatively uncontaminated reference area. This was taken to indicate an *indirect* effect of pollution, causing reduced immune-system

strength in the eels and thus higher vulnerability to other disorders even if these had not shown at the time of sampling and even if they might be finally induced by a separate natural factor. Focusing on a single-variable direct-cause explanation would in such a case completely miss the damaging role of pollution, and put the effects down to 'natural causes';

3 therefore that, with due caution, circumstantial evidence for cause–effect mechanisms is legitimate.

Indeed, on closer inspection all scientific reasoning is unavoidably circumstantial as the radiocaesium soil example also illustrated. Returning to the competing idioms of physical supercomputer modelling methodology, and broader-based, historical climatology, the charge that the latter employed only surrogate variables and hence could not gain intellectual control of its inferences and causal constructs can also be seen to be misleading. The mathematical models also have to employ surrogate parameters and variables for the real values and behaviours of interest; it is just that these are more culturally esoteric, and more socially and intellectually inaccessible, so that this characteristic is not so evident as it was for the geographers and others in their research. The models may also be more buttressed by cross-cutting and interlocking scientific constructs in a 'thicker web' of consolidation than the less reductionist scientific idioms.

The conventional assimilative capacity scientific idiom of marine pollution (Campbell and Chadwick 1993) appears on the face of it to avoid circumstantial reasoning; but for example, assimilative capacity scientists claimed to have disproven the alleged connection between fish disease and contamination when the observation of high levels of disease away from inshore waters was reported. This was an anomaly sufficient to 'disprove' the connection, on the prevailing assumption that such offshore waters are less contaminated. This conclusion was drawn before measurements which showed the 'offshore is cleaner' assumption to be wrong – at least in this case. The convinced application of the general assumption (that offshore waters are cleaner) to the specific case was just as circumstantial as was the reasoning sometimes adopted by the 'precautionary' idiom.

There is always an ineradicable element of indeterminacy in deciding whether a new empirical situation is an instance of a class of entities under one theory or model, or another. (Is the soil in the upland sheep areas the same or different from the soil(s) on which the conceptual model of caesium behaviour is constructed? – it depends on whether we are concerned about sheep meat contamination or direct external gamma-ray exposures.) The traces of this endemic indeterminacy are usually already well-concealed (even from the scientists involved) by the time it comes to exercising policy responsibilities, even though the way the choices are made at such scientific points may have important policy implications.

We have learnt from the detailed analysis of the creation of scientific knowledge over the last twenty years or so that many of the intellectual commitments which constitute that knowledge are not completely validated, not fully determined by empirical nature (Collins 1985, Latour and Woolgar 1979, Barnes and Edge 1982). Always central to the process are not just uncertainties in the form of imprecision (which, it is assumed, will be narrowed down by more research), but fundamental indeterminacies – for example, as to whether things are classified as the same or different, and on what specific properties or criteria. The purely technical aspects of such intellectual commitment merge with epistemic questions as to why we are constructing such knowledge anyway; this is always open to social evaluation and negotiation, though that is very far from saying that scientific truth can be subject to social choice.

However, we can see that when scientific knowledge is deployed in the public domain, the social judgements of a relatively private research community which create closure and 'natural validation' around particular constructions of speciality scientific knowledge, need to be re-opened (deconstructed) and renegotiated in a wider social circle: possibly one involving different epistemological commitments and expectations, and correspondingly different definitions of the boundaries between nature and culture, or (objective) determinism and (human) responsibility. These too will have to be recognized and renegotiated in some way, as new, more-broadly legitimated principles on which scientific knowledge generation can be founded.

CONCLUSIONS

This chapter has attempted to outline the basis of a fundamentally different framework for thinking about the relationship between scientific knowledge and public policy, such as exists around global environmental issues. In particular it focuses upon the complex question of authority and credibility. It suggests that these should be conceptualized more in terms of social and cultural identification (or alienation) than in terms of the intensification of the natural warrant for policies, and then in their successful communication. These problems of cultural identification are more fundamentally difficult when the ambitions of scientific policy enterprises are inflated to global level; but, even within the cultural confines of the 'scientific' industrialized world, they are already more deeply problematic than is usually imagined.

The analysis shows that the construction of scientific knowledge is less completely determined by nature than conventional approaches assume, and that as a result the construction processes are more open-ended and contingent than is usually recognized. Sociological analysis of scientific knowlege has shown that the successful 'closure', and stabilization of these open-ended, incomplete constructs or knowledges, is achieved by mutual overlap and the reinforcement of provisional or incompletely warranted commitments by adjacent social–intellectual networks. An incompletely warranted intellectual commitment in social group A is 'confirmed' by its consistency with an apparently solid, but just as incompletely warranted, commitment in social group B; and vice versa. Thus both are consolidated and 'closed' as if naturally determined by processes of mutual affirmation.

This kind of network interdependency and mutual 'bootstrapping' of credibility is well-recognized in the sociology of scientific research, via interaction amongst different research specialities and groups. However it also occurs between 'scientific' and 'policy' networks and actors, in the myriad intermediate roles and institutions such a advisory committees, review working parties and the related institutional flux. Representations of the boundaries of science with policy are flexibly – and often inconsistently – articulated as an integral part of these processes.

As the earlier case-studies showed, one result is that social commitments, whether consciously or inadvertently, are built into the 'natural' knowledge so constructed and deployed as policy interventions. One might say that this shows science responding to changing dominant social values, in something akin to a respectable model of democracy. However, especially in the light of science's historical role as an ideological resource for dominant interests, this would be an extravagantly optimistic interpretation of the capacity of such processes as presently institutionalized, not only to reflect democratic values, whatever they are, but to encourage their mature and more effective articulation in relation to science and environmental challenges.

At this point it is important to distinguish between two typical approaches from within the sociology of scientific knowledge, because they carry very different implications in this context.

An interests-based approach would argue that scientific knowledge is shaped by the outcome of struggles and negotiations between different social interests, including the social-cognitive interests which scientists committed to a particular paradigm in reproducing and extending their familiar forms of technical practice and identity. Thus 'internal' and 'external' social interests may combine to produce particular bodies of 'natural' knowledge in which the essential indeterminacies referred to above are closed by those interests.

This approach would argue that dominant scientific knowledge, as constructed in the supercomputer climate models used by the IPCC, reflects developed countries' interests in obscuring the social and political inequities which lie at the heart of global environmental degradation. Buttel and Taylor (Chapter 11, this volume) appear to express this perspective in advocating a sociology of science for global environment. A similar 'interests' analysis was offered in the early 1970s about the systems computer modelling of global environmental disaster in *The Limits to Growth* (Golub and Townsend 1977).

However, Buttel and Taylor's persuasive general argument for a sociology of scientific knowledge approach to global environment does not make clear the distinction between their assumed analytical perspective and a more 'cultural' approach. In Chapter II they rightly note that

> when sociologists have attended to global change issues, they have tended to do so by uncritically accepting and appropriating the global 'constructions' of modern environmental problems that have emerged within both the environmental sciences and the environmental movement (p. 228).

They then observe that

> This is especially problematic since, within both science and politics, the 'globalization' of the environment has served to steer attention to common human interests in environmental conservation, and away from analysing the difficult politics that result from different social groups and nations having highly variegated – if not conflicting – interests in contributing to and alleviating environmental problems (pp. 228–9).

They then point out that such sociologists, by uncritically swallowing global environmental science constructs, have been caught unawares by the force of the radical critique of 'global environmentalism' and its rich world agenda, now emanating from the developing world. In other words, sociological deconstruction of global environmental science is necessary to expose the hidden interests of the rich countries in concealing the more demanding political challenges of exploitation, domination and inequity underlying global environmental change.

This kind of analysis is easy to accept. However, in the light of the cases described before, and indeed of much other work in sociology, the question is whether it goes far enough. It leaves largely untouched the possible connections and contingent reinforcements between scientific knowledge, constructions of the human subject, and the cultural milieu of late-modern society.

A different emphasis in sociology of global environmental scientific knowledge would also consider the ways in which scientific knowledge of environmental situations naturalizes and reinforces particular cultural and moral values or identities beyond the reach of what can be called 'interests'. Thus the inadvertent structural features of scientific knowledge which, for example,

- classify natural and social 'actors' in one way and not another;
- homogenize and 'fix' them at a particular scale of aggregation;
- define ambivalence as antipathetic to rationality because it undermines the assumed epistemological principle of prediction and control;
- rule circumstantial reasoning in one direction to be unsound, whilst covertly if innocently using it to construct alternative 'natural' conclusions; or
- define human actors as interested only in maximizing or satisficing utilities, and conceive the social science of global environment in such terms;

can be seen to reflect and to naturalize particular cultural norms, social identities, and even models of human nature which perhaps need to be problematized as part and parcel of the global environment predicament.

To give an example, unreflexively adopted implicit models of human agency buried within, and acting as preanalytical framing commitments that shape global environment scientific knowledge, may produce unpleasant surprises for policy-makers. The assumption is that increasing public awareness of global warming scientific scenarios will increase their readiness to make sacrifices to achieve remedial goals. Yet an equally plausible suggestion is that the more that people are convinced that global warming poses a global threat, the more paralysed they may become as the scenarios take on the mythic role of a new 'end of the world' cultural narrative. Which way this turns out may depend upon the tacit senses of agency which people have of themselves in society. The more global this context the less this may become. Thus the cultural and social models shaping and buried within our sciences, natural and social, need to be explicated and critically debated.

The unreflexive nature of science, including a substantial part of social science in this arena, serves to avoid the problematization, for example, of social relations where instrumental epistemologies derived unwittingly from the natural sciences assume essentialist or rational

186

choice models of social behaviour. This imposes the prescriptive assumption that social interaction is not an activity with moral meaning and struggle against indeterminacy, to create value and identity, not an end in itself, but merely a means to the supposedly ultimate ends of maximizing utilities or preferences.

Likewise, a more culturally oriented sociology of scientific knowledge would recognize more of the reflexive questions to be asked about the intrinsically alienating effects of knowledge which positions and constructs people in environmental processes as if they are merely reproducing and extending consumer-based capitalism. Leslie Sklair, in Chapter 10, comments that

> For too long the social study of global environmental change has been focused on supply-side issues, (predominantly production systems), and has paid too little attention to demand-side issues (particularly consumption patterns) (pp. 221–2).

This comment resonates with the simple observation that for global sustainability rich-world, consumption-oriented behaviour is going to have to change more radically than is implied in the language of the economic and technical adjustments which monopolize mainstream policy thought. The changes are likely to bite much deeper, into the very social relations and identities from which we can gain sustenance (which is arguably vicariously sought from consuming material resources and generating physical environmental impacts).

However much or little this is true, the practical large-scale social exploration of these dimensions is already underway according to the sociological analysis of new social movements. The sociology of scientific knowledge is essential as an intellectual resource to retrieve the domains of human struggle, cultural reconstruction and social responsibility blindly sequestered by science, including its social science counterparts.

A final comment relates to the familiar criticism that a 'relativist' sociology of science undermines the basis of political critique, of the hidden interests shaping science as an ideological tool which needs debunking. The simple interests approach identified above stops short of that reflexivity, and in so doing retains the realist stance from which the radical critique of dominant versions of global environmental science can be mounted. It is notable that much of that critique (see, for example, Shiva 1989) displays a markedly realist tone.

So does the more reflexive version of the sociology of scientific knowledge – that which problematizes the epistemological commitments, cultural resonances and inadvertent social models implicit within scientific constructions – necessarily undermine critique? I would conclude that it does not; indeed that the reflexive critical

BRIAN WYNNE

examination which it helps to provide of late-modern society's deeper identifications with modern science as (instrumental and standardizing) culture is an essential component of an authentic and constructive response to global change in its widest sense. As such, and even though the relationship remains to be worked out, it ought to be a complementary perspective to the less reflexive, more directly political critique of the global environment problematique as defined by science, and of which IPCC may be considered as a modernist delusion – not so much in terms of its relationship to physical realities, whatever they may be, but in terms of its apparent expectations of cultural authority and global social purchase.

The sociology of global environmental science indicated here would imply the reopening of explicit and diverse negotiations of pluralist epistemological commitments within a more open-textured, culturally differentiated and socially permeable 'scientific' realm. This would inevitably mean that policy cohesion, presently vainly pursued through the ever more anxious and manifest manipulation of scientific consensus and 'natural authority', would have to be more openly based upon human commitments in the face of recognized indeterminacies. This reopening of such a human domain is not without its own risks. But it would reinforce the stance taken by many in the field that the policies most likely to help in combating global warming are worth doing anyway, on social, political, moral and even economic grounds, regardless of what they may or may not do to the environment.

This focus on the implicit cultural framing of scientific knowledge does not mean that such knowledge would be debunked or denied authority. Rather the conditions of validity would be critically explored, and the tacit social and moral commitments of knowledge exposed for debate and negotiation. This would demand the negotiation of plural forms of science with negotiable social and cultural boundaries, and correspondingly more social struggle to articulate emergent values and fluid identities. In this kind of process scientific uncertainties would not be an embarrassment, but – seen more properly as authentic human indeterminacies – the meat and drink of a more mature social learning process.

REFERENCES

ASCEND (1992) International Council of Scientific Unions, *An Agenda of Science for Environment and Development into the 21st Century*, Cambridge and New York: Cambridge University Press.
Barnes, S.B. and Edge, D.O. (eds) (1982) *Science in Context: Selected Readings in Sociology of Science* Milton Keynes: Open University Press.

Beverton, R. (1992) Former Secretary, UK Natural Environment Research Council, personal communication.

Brundtland Commission (1987) *Our Common Future*, UN World Commission on Environment and Development, Oxford: Oxford University Press.

Campbell, J. and Chadwick, M. (1993) 'Assimilative capacity and critical loads as preventive environmental science-policy instruments', in T. Jackson (ed.), *Towards Cleaner Production*, Stockholm: Stockholm Environment Institute/Lancaster University CSEC.

Collins, H. (1985) *Changing Order: Replication and Induction in Scientific Practice*, London and Beverly Hills: Sage.

Dethlevsen, V. (1988) 'Assessment of data on fish-diseases', pp. 276–85 in P. Newman and A. Agg (eds), *Environmental Protection of the North Sea*, London: Heinemann.

Gale, H.J., Humphreys, D.L. and Fisher, E.M. (1964) 'Weathering of caesium-137 in soils', *Nature*, no. 4916, 18 Jan., 257–61.

Golub, R. and Townsend, J. (1977) 'Malthus, multinationals and The Club of Rome', *Social Studies of Science* 7: pp. 201–222.

Jasanoff, S. (1990) *The Fifth Branch: Science Advisers as Policymakers*, Cambridge, Mass.: Harvard University Press.

Latour, B. and Woolgar, S. (1979) *Laboratory Life* (2nd edn 1985), London: Sage.

Liverman, D. (1991) 'A case-study of policy learning and global warming in Mexico', Draft paper for the project, Pennsylvania State University, Geography Dept., Social learning and global environmental issues.

Lunde, L. (1991) *Scientific Knowledge and Policymaking. The Case of IPCC*, mimeo, Oslo: Fritshoff Nansen Institute.

Maddox, J. (1972) *The Doomsday Syndrome*, London: Robbins.

Mason, B.J. (1976) 'Towards the understanding and prediction of climatic variations', *Quarterly Journal of the Royal Meteorological Society* 102 (433), July: 473–98.

Nelkin, D. (ed.) (1979) *Controversy: The Politics of Technical Choice*, London: Sage.

Ravetz, J. (1990) *The Merger of Knowledge with Power*, New York: Mansell.

Robbins, D. and Johnson, R. (1976) 'The role of cognitive and occupational differentiation in scientific controversies', *Social Studies of Science* 6: 163–92.

Ross, A. (1991) *Strange Weather: Culture, Science and Technology in an Age of Limits*, London: Verso.

Shiva, V. (1989) *Staying Alive: Women, ecology and development*, London: Zed Books.

Sindermann, C.J. (1984) 'Fish and environmental impacts', *Archiven die Fische Wissenschaft* 35(1): 125–60.

Smithson, M. (1989) *Ignorance and Uncertainty: Emerging Paradigms*, Berlin and New York: Springer Verlag.

UK White Paper (1990) *This Common Inheritance: Britain's Environmental Strategy*, London: HMSO.

Wynne, B. (1991) 'Misunderstood misunderstanding: social identities and the public uptake of science', *Public Understanding of Science* 1(3): 281–304.

9

FACING GLOBAL WARMING

The interactions between science and policy-making in the European Community

Angela Liberatore

INTRODUCTION

Scientific uncertainties and political controversies surround the issue of the risk of global warming due to anthropogenic sources. Since global warming reached the international and the EC political agendas, scientific findings and uncertainties are used as arguments for deciding between different courses of action (mainly between taking precautionary action or postponing decisions) and selecting policy instruments.

The specific features of the interactions between science (natural and social sciences) and policy-making in the EC context are analysed in this chapter to explain, on the one hand, how and why science and experts' advice shape EC policy with respect to the global warming issue and, on the other, to see whether and why scientific developments are influenced by political needs.

With respect to the first point, it is the transformation of scientific findings into policy issues which is focused on. With respect to the second point it is noted that previously 'marginal' research becomes 'mainstream' when its contribution is required for 'high politics'.

THE EMERGING OF THE GREENHOUSE ISSUE IN THE EUROPEAN COMMUNITY[1]

The risk of human-induced climate change was first addressed at the EC level as a research issue. This occurred in December 1979 when the Council Decision to adopt a multiannual EC research programme in the field of climatology (see EC Commission 1980) addressed the need to better understand man–climate interactions and forecast possible climate changes and their impacts.

This decision was the result of internal elements like the existence of EC research programmes and organizations in the field of the environment,[2] and to an external event such as the drought of 1976 which seriously affected some African and European regions.

The need to understand whether phenomena like the drought of 1976 were due to anthropogenic sources was in fact used by members of DG XII of the EC Commission as an argument for starting EC research in the field of climatology within the framework of the EC research programme on environment. The suggestion to start EC research on climatology was initially opposed by some member states that did not see the need for an EC effort in such a field. However, this opposition was overcome due to the emphasis on the regional and global dimension of climate change put forward by members of DG XII and a group of experts appointed to draft the new programme, and due to the support to the Commission's initiative by some small countries interested in joining forces in environmental research and by Southern countries alarmed by the issue/risk of desertification.[3]

For almost ten years, from the adoption of the first research programme in the field of climatology until 1988 (when a policy document on the greenhouse issue was drafted by a Commission's interservice group), the greenhouse effect and climate change were regarded only as scientific issues by the EC Commission, the activities of which were restricted to the promotion of relevant research. The small budget initially allocated to such research (8 million ECU for the first five-year sub-programme on climatology out of 42 million for the whole environment programme, and 17 million for the second sub-programme from a total of 75 million) indicates that, beside being regarded as a 'merely' scientific issue, climate change was not considered a research priority by the Commission.[4] Moreover, until the approval of the environment programme 1991–4, EC-sponsored research was exclusively devoted to the natural sciences, social, economic and political aspects.[5]

The aim of the climatological research was to reach a better understanding of the physico-chemical processes relating to the increasing concentrations of greenhouse gases in the atmosphere and to predict the corresponding climate changes and their possible impacts. A symposium on these themes[6] was organized by DG XII in November 1986 and represented a major EC effort to improve the networking among leading European and US scientists in the field, at the same time providing a high-level review of the state of available knowledge. The existing uncertainties about elements like the timing and regional impacts of climate change (although climate change itself was assessed to be certain) were stressed, and the main policy recommendation offered at the symposium regarded the intensification of research to reduce those uncertainties. Some policy actions to prevent or mitigate the possible effects of climate change were also suggested by some participants (for example, the reduction of F-11 and F-12 production, and the development of a 'resilient' energy system), but at that time they remained confined to scientific circles. Another recommendation, which addressed the major problem of the gap between science and policy-making, regarded the exchange of information and the collaboration between scientists and decision-makers.

EC research in climatology provided an important background element when global warming became a policy issue, since it made scientific findings known and expressions like 'greenhouse effect' and 'climate change' household words which were bound to reach the ears of EC policy-makers.[7] However, it did not provide a direct input for the starting of the EC greenhouse policy. In other words, the observed sequence does not involve a causal relation between EC research and EC policy with respect to the greenhouse issue.

While some references to the symposium on 'CO_2 and other greenhouse gases' appear in the Commission's 'Communication to the Council on the greenhouse effect and the Community' (COM (88) 656 final), the results of EC research in the climatology field are hardly referred to in other EC policy documents on the greenhouse issue. On the one hand, this is due to the general features of EC research programmes: that is, to the fact that while they perform an important catalytic role by sponsoring collaboration between European scientists they do not yet represent (at least in our case) as authoritative a scientific forum as do other international or intergovernmental programmes. On the other hand, until the early 1990s EC research relating to the climate change issue did not address the technological, social, economic and political aspects relevant for guiding policy-making. Research on these aspects (for example, about the cost-effectiveness of energy efficiency measures and their impact on CO_2 emissions) started being addressed in 1990 (mainly within the research programme on non-nuclear energy, JOULE, of DG XII; then by studies of DG XVII, Directorate-General for Energy; and DG II, Directorate-General for Economic Affairs). It is worth noting that those studies did not influence the emergence of the greenhouse effect as a policy issue, but were rather a result of such 'policization'.[8] In turn, as we shall see (pp. 196–7), they contributed in shaping the response assessment and the formulation of EC strategy to deal with climate change.

The framing of the greenhouse effect as a policy issue – rather than as a scientific issue alone – to be dealt with at the EC level started in 1986 with two reports prepared by members of the European Parliament. In July of that year a report by Mrs Bloch von Blottnitz on renewable energy sources pointed out that they do not produce

CO_2 emissions, and a report on the greenhouse effect by Mr Fitzsimons acknowledged the existing scientific uncertainties but also argued for taking some immediate countermeasures in the fields of industrial, agricultural and energy policy, and in the production of propellants. These reports can be regarded as anticipating some features (like the focus on energy efficiency) of the EC 'greenhouse policy'; however, they did not play a direct role in starting and shaping such a policy. The 'real' beginning came more than two years later and was mainly developed by the Commission, since the European Parliament has only a consultative role in environmental, energy, fiscal and other ('greenhouse relevant') policies.

In 1987, the greenhouse issue was mentioned by the EC Commission in its draft resolution regarding the fourth environmental action programme. This document states that,

> Looking further ahead into the future it is clear that difficult problems could arise from the use of fossil fuels if the build-up of atmospheric CO_2 and the 'greenhouse effect' are shown (as certain scientists fear) to have serious impacts on climate and agricultural productivity worldwide. In case further research should confirm the likelihood of such impacts, the Community should already be thinking about possible responses and alternative energy strategies.[9]

Interestingly, the reference to possible responses and alternative energy strategies is not included in the final text of the fourth environment action programme,[10] where the greenhouse effect is only referred to as a topic of the EC research programme in climatology and natural hazards. This is an indication of the EC Commission's reticence at that time to regard the greenhouse effect as anything other than a scientific issue.

However, a few months later – in July 1988 – the Commission decided to establish an 'interservice group' (that is, a group formed by members of various Directorate-Generals) to examine the problems relating to the greenhouse effect. The creation of such a group meant that the Commission had decided to devote more political attention to the greenhouse effect and to explore the possibility of making proposals to the Council on that issue.

Beside the ongoing internal debate (within the Commission and the Parliament) on the greenhouse effect, an external event that played a crucial role in this change of attitude and the corresponding EC 'policization' of the greenhouse issue was the Toronto Conference on 'The changing atmosphere'. Implications for global security' held in June 1988. In the previously mentioned 'Communication to the Council on the greenhouse effect and the Community' prepared by the interservice group (with DG XI, the Directorate-General for Environment, taking the main responsibility for it and for the later developments of the EC greenhouse policy), emphasis is put on the initiatives recommended at the Toronto Conference – defined as a 'particularly important event' – such as the 20 per cent reduction of CO_2 emissions in industrialized countries, the encouragement of energy policies aimed at reducing CO_2 emissions, the ratification of the Montreal Protocol, and the promotion of a global convention on climatic change. Moreover, the related future steps of international action, from early 1989 until the 1992 United Nations Conference on Environment and Development, are mentioned. In other words, international attention towards the greenhouse issue is presented in the document as the context and stimulus for starting an EC greenhouse policy.

Since the Communication of 1988, the greenhouse issue remains permanently on the Commission's policy agenda, and with increasing importance. In the years that followed, the 'greenhouse policy' developed as a transversal policy related to the energy, economic, fiscal and external relations policies, and to the research and environmental policy.

Regarding this first phase of the greenhouse effect as a policy issue to be dealt with at the EC level, some points deserve attention with respect to the interaction between science and policy-making.

Natural science findings provided the basis for identifying the problem and framing climate change as a global, largely human-induced and very serious threat. It is not by chance that the Commission's 'Communication on the greenhouse effect and the Community' starts with a relatively long and detailed description of the available knowledge and existing uncertainties regarding the effect.

It is worth mentioning that the scientific assessments reported in that document were mainly 'imported' from US sources (such as the Department of Energy, the Environmental Protection Agency, the National Research Council, the World Resource Institute).[11] While some conclusions of the EC symposium on 'CO_2 and other greenhouse gases' are referred to, European sources of assessment were less extensively quoted and almost exclusively so with respect to the issue of the possible impacts of global warming.

The overall scientific assessment referred to by the Commission as a basis for evaluating alternative courses of action can be summarized as follows:

1 the main contributor to the greenhouse effect is CO_2;
2 the main source of CO_2 is fossil fuel combustion;
3 the available general circulation models indicate that global

temperature increase (between 1.5 and 4.5 degrees) will necessarily occur if atmospheric concentrations double with respect to pre-industrial concentrations;

4 the available models are not able to assess with confidence the possible regional increases in temperature but some studies indicate that in Europe these increases could be above the world average;

5 climate change may cause various negative impacts (sea-level rise, depletion of water resources and agricultural production, health problems, etc.) not homogeneously distributed.

On the basis of such an assessment (which was then refined, especially in the light of the report of the Intergovernmental Panel on Climate Change, but not radically changed in the following years), a first 'screening' of possible responses is offered in the Commission's Communication. Such screening (which was largely influenced by the conclusions of international conferences like the Toronto, Villach and Bellagio Conferences)[12] includes precautionary measures in the fields of energy and forestry, adaptive measures regarding agriculture and sea-level rise, and other possible actions at the international level such as the reduction and ban of CFCs.

TRANSFORMING CLIMATOLOGICAL FINDINGS INTO POLICY ISSUES

The above mentioned 'screening' of possible responses points to a very important aspect of the interactions between science and policy-making.

Natural science findings about increased concentrations of certain gases in the atmosphere and their contribution to the production of the so-called greenhouse effect needed some 'translation'[13] in order to become policy issues. This required the identification, evaluation and selection of measures and strategies aimed at preventing or mitigating the possible negative impacts of global warming. In other words, to become a policy issue the greenhouse effect had to become a 'treatable' problem, even if an extremely difficult, though a politically 'manageable' one. In this respect, the role of economists, policy analysts and energy technology experts has been crucial and it is their contribution – rather than that of natural scientists – which now represents the main scientific input to policy-making in the EC context.

Natural science findings still play an important role in the international debate about taking precautionary measures or postponing action. The existing uncertainties about physical processes (like ocean-atmosphere interactions and other aspects), about the size and timing of temperature increase, and about the distribution of possible impacts are used either to argue (as the Bush Administration did) that such uncertainties must be reduced before taking action, or to argue (as in the case of the European Community, the Scandinavian countries and others) that precautionary measures must be designed on the basis of available knowledge and uncertainties to avoid being too late. However, in this debate the attention also seems to be focused – more on the uncertainties regarding the costs of precautionary and adaptive actions than on the uncertainties regarding climatic processes – especially following the UN Conference on Environment and Development.

In any case, especially when the precautionary approach is favoured and the decision to take action is made (as in the EC case), socio-economic and technical experts are those whose advice is looked for in the policy process.

Coming back to the 'Communication on the greenhouse effect and the Community', it can be seen that the whole spectrum of precautionary and adaptive measures is presented. In later documents of the EC Commission a more selective approach can be observed, since from the 1990 proposal for 'A Community action programme to limit EC carbon dioxide emissions and to improve the security of energy supply',[14] the Commission has started to focus on the energy sector and on energy-related measures (including energy efficiency standards, energy/CO_2 tax, etc.).[15]

This was partly the result of previous policies, which provided the 'filter' for selecting possible measures and fields/sectors for action, and of economic-, policy- and technology-oriented research.

As far as the first element is concerned, EC environmental policy, and especially its focus (since the fourth environmental action programme and the Single European Act) on the need for preventive action and for integrating environmental concerns in all policy sectors,[16] provided some criteria for addressing the greenhouse issue as a problem that required anticipatory and cross-sectoral policy action. Moreover, EC energy policy – especially since 1986 (with the so-called 1995 objectives) – has emphasized the goal of improving energy security and energy efficiency,[17] and provided some instruments (energy efficiency measures) for a greenhouse policy which is not only environmentally beneficial but also economically so. Also, external relations policy, particularly EC willingness to take a leading role in the international arena (in the environmental and in other fields),[18] favours the selection of both visible and feasible policy measures to show EC ability to translate declared commitments into practical actions.

With respect to economic-, policy- and technology-oriented research, this is partially promoted by DG XII (for example, the JOULE

196

programme on non-nuclear energy and its sub-programme CRASH on cost-efficient policy options for CO_2 emission control) and partially by other Directorates-General. One can mention, for instance, the study by DG XVII *Energy for a New Century* (Energy in Europe 1990), the DG II study *The Economics of Policies to Stabilize or Reduce Greenhouse Gas Emissions: The Case of CO_2* (Mors 1991), and the study commissioned by DG XI *The Development of a Framework for the Evaluation of Policy Options to Deal with the Greenhouse Effect* (CRU and ERL 1990). These studies and programmes can be regarded as the economic, policy and techno-logical 'translations' needed to transform climatological findings into policy issues.

These 'translations', together with the IPCC assessment concerning some basic characteristics of the greenhouse gases,[19] have been used in the identification and selection of policy options – particularly in guiding or simply supporting (the main criteria and reasons for such a selection are not necessarily scientific ones) the decision to focus on certain sectors, like energy, and certain policy instruments (economic incentives, fiscal instruments, regulatory measures).

Concerning this last point, one can take the case of the debate on the CO_2 energy tax to see how important has been the economic and technical 'translation' of climatological findings to select for attention a specific policy instrument.

Beside political arguments about the desirability for the Community to take a leading role in international negotiations on the climate change issue (especially advocated by the former Commissioner for Environment, C. Ripa di Meana), economic and technical exper-tise played an important role in selecting for attention and designing the specific characteristics of the CO_2 energy tax. In particular, the previously mentioned studies promoted by DG XII on the cost-effectiveness of CO_2 reduction options argued for the existence of a remarkable energy saving potential and the possibility to exploit such potential through a tax (Capros *et al.* 1991, Coherence 1991); this issue was also analysed in the DG XVII study (see Energy in Europe 1990). Moreover, DG XII studies and the DG II study by Mors (1991) provided an economic basis for analysing the possible macroeconomic impacts of a CO_2 energy tax and for formulating specific features (like its revenue neutrality and the amount) of such a tax.

In general, the above-mentioned political arguments (that is, the theme of international leadership), together with the economic and technical 'translations' of climatological findings, provided the ground for the moderately precautionary strategy of the Community, the so-called 'no-regret' strategy.[20]

197

POLICY NEEDS AND SCIENTIFIC DEVELOPMENTS

An increased emphasis on the role of R&D can be found in the most recent documents of the EC Commission, not only (and not mainly) for the understanding and the reduction of uncertainty regarding the greenhouse problem, or for the economic as well as the policy 'translation' of natural science findings, but especially concerning the implementation of the selected options.

An interesting indicator of the increased support to such types of R&D activities as part of the EC strategy is the funding allocated for the THERMIE programme on the promotion of energy technologies in Europe: 700 million ECU.[21] This is more than the 551 million ECU funding allocated to the nuclear fusion programme,[22] which is the leading EC research programme after the one on information technologies, ESPRIT. Moreover a specific action and research programme, SAVE (Specific Actions for Vigorous Energy Efficiency), has been formulated to influence the behaviour of consumers and producers in that field.[23] Another action programme, ALTENER (Alternative Energies), has been proposed which should cover provisions designed to remove barriers to the marketing of new and renewable energy sources.[24]

These data indicate how a previously marginal research field such as research in energy efficiency and renewable energy sources become 'mainstream' research when it agrees with 'high politics' purposes.

After being advocated for decades, but with few results, by environmental NGOs and green parties as an alternative to nuclear power and other risky and/or polluting energy sources, and after a short appearance in Western countries' energy policy agendas following the oil crisis of 1973, the research on and introduction of energy efficiency and renewable energies are now a top priority of EC research, energy and environmental policy.

While not being exclusively linked with the greenhouse issue, this shift is largely due to the fact that energy efficiency and renewable energies provide a politically and economically palatable tool for the EC greenhouse policy; a policy which 'crosses' crucial sectoral policies such as economic, fiscal and external relation policies, beside research, energy and environmental policies.

The funding for, and importance attributed to, basic research relevant for the global warming issue also increased in the meantime, although the increase wasn't as great as in the previously mentioned technology-oriented research.

In 1989[25] the programme EPOCH (European Programme on Climatology and Natural Hazards) was given 40 million ECU[26] of the overall environment programme. Moreover, in 1989 50 million ECU were allocated to the newly introduced MAST programme (Marine Science and Technology), which includes some research relevant for the greenhouse issue, like ocean–atmosphere interactions and technologies that could be used to face sea-level rise.

More recently, the importance of improving knowledge on the greenhouse problem and of contributing to international programmes on global change is emphasized in the general objectives of the EC environment programme 1991–4, where research related to climate change is grouped in the area 'Participation in global change programmes'. Research on global change is also included in the 1992–4 activities of the Environment Institute of the EC Joint Research Centre (EI 1991). Even if this represents more of a new 'label' than a new field of research of the JRC,[27] it indicates the increased importance of climate change research among EC research and policy priorities.

The same can be said, at least in part, regarding the initiative to start closer collaboration with the European Space Agency (ESA) in the field of earth observation. While this is part of an increased realization of the importance of space research and technology for many sectors of direct interest of the EC (security, information technologies, trade, agriculture, environment), it is interesting to note that earth observation is regarded as the most important of future space activities for the EC.[28] This activity (which includes the launching of ESA's satellite ERS-1 and the preparation of ERS-2) is directly related to the greenhouse issue (see, ESA 1988, 1991). Although this attention to satellite observations is not new in the 'landscape' of EC research, since remote sensing activities had already been conducted by the JRC, it indicates a new emphasis – an emphasis which could involve the use of the greenhouse issue as a means to support 'big science' (or at least 'big technology') research in a field which is different from the 'classic' one of nuclear fusion.

CONCLUSIONS AND PERSPECTIVES

The main aspects of the complex interactions between science and policy-making in the EC context, and with specific reference to the global warming issues, can be summarized as follows.

Natural science research (especially climatological) played an important role in the identification and assessment of the risk of global warming. Such identification and assessment represented an important basis for the emergence of the greenhouse effect as a later policy issue. Due to specific features of the EC research policy and institutional setting, the scientific assessments referred to in policy documents were/are mainly imported from US (DOE, NRC, etc.) and international (like the IPCC) sources.

To become policy issues, climatological findings needed a 'translation' able to make the risk of global warming a 'treatable' and policy 'manageable' problem, albeit an extremely difficult one. Such a translation was provided by previous policies and by research on economic, policy and technological aspects. Previous policies provided the 'frames' to look at the greenhouse effect in terms of energy, environmental, external relations and other policy goals; and research on economic, policy and technological aspects helped policy-makers to design and select policy instruments aimed at tackling the greenhouse problem.

Research on these aspects becomes especially important when the decision is made (as in the EC case) to take some precautionary action in spite of (or better, on the basis of) the existing scientific uncertainties. In turn the development of such research is being favoured by the 'policization' of the global warming issue. For instance, the previously marginal research on energy efficiency and renewable energies is now a top EC research priority since it matches the 'no-regrets' greenhouse policy by providing economically (and environmentally) beneficial instruments. Moreover, basic research on climate change is also being emphasized more, and this could lead to the reduction of existing uncertainties (but also the production of new ones) and/or to the identification of new threats. Finally, the greenhouse issue may favour the development of a new kind of 'big science' (or 'big technology') space research.

The trend towards and increase of EC interest in and funding for global change research activities is likely to continue in spite of the turmoil surrounding the Maastricht Treaty and the emphasis on the 'subsidiarity principle'.[29] While such a principle is advocated by 'Euro-sceptics', to return to the member states the policy competences in various sectors (including environmental policy and research policy) which are presently also covered by the EC cannot but favour the implementation of Community action in relation to the global warming issue.

Both in terms of 'scale' and 'effects', research and environmental policies aimed at addressing such an issue (and other global or regional environmental problems) are necessarily 'transboundary' ones. That is, while requiring action at the national level (especially in areas such as the implementation of environmental measures or scientific training and education), those policies must be developed at a supranational level in order to face problems which disregard national boundaries. In this respect, policy actions and 'big science' activities in the field of global change can be better pursued – consistently with the still vague but influential subsidiarity principle – at the Community level.

NOTES

1 The reconstruction of the EC greenhouse policy is based on work conducted within the framework of the international project on 'Social learning in the management of global environmental risks', directed by W. Clark, Harvard University, and partially supported by the John and Catherine MacArthur Foundation.

2 EC research includes both 'direct' research conducted within the EC Joint Research Centre (JRC) and 'indirect' research: that is, research funded partially by DG XII (Directorate-General for Science, Research and Development) of the EC Commission, and partially by member states' research institutions. The latter represents the main component (in terms of budget and scope of research) of EC research, and it is the one referred to in the text when not specified otherwise. The first environmental research programme was launched in 1976. On EC environmental research see Liberatore (1989).

3 A symposium organized by DG XII on the desertification issue was held some years later, in 1984, in Greece. The connections between such issue and the possibility of man-induced climate change due to increasing atmospheric CO_2 is mentioned in the 'Introduction' of the *Proceedings* (see Fantechi and Margaris 1986) and in some papers presented at the symposium.

4 In this respect it must be added that environmental research in general is not regarded as an EC research priority. Top research priorities are the development of information technologies and energy – especially nuclear fusion – technologies.

5 The 1991–4 environment programme includes a research area on economic and social aspects of the environmental issue. This was the result of pressure from the European Parliament and interest in socio-economic aspects by top level members of the Commission, including the Director-General of DG XII, the Commissioner of Research, and the President of the Commission.

6 The proceedings were edited by Fantechi and Ghazi (1989).

7 This circulation occurred (and occurs in general) because the research programmes promoted by DG XII have to be presented to other DGs, and must be approved by the Council. Moreover, the symposium on 'CO_2 and other greenhouse gases' (see Fantechi and Ghazi 1989) was strongly supported by the Vice-President of the Commission, at that time K.H. Narjes.

8 The same can be said for the increased funding for basic research related to the global warming issue. On this point see pp. 188–9 of this chapter.

9 OJ C 70, 18 March 1987, p. 13.

10 OJ C 328, 7 December 1987.

11 The main US studies quoted in the Communication are the following: DOE 1985, DOE 1988, EPA 1986, NRC 1987, WRI 1987.

12 Conference on the Assessment of the Role of CO_2 and other Greenhouse Gases in Climate Variations and other Associated Impacts, Villach, 1985 (sponsored by WMO, UNEP and ICSU); Conference on Developing Policies for Responding to Climate Change, Bellagio, 1987 (sponsored by WMO and UNEP).

13 On the issue of the 'translation of climatological facts into political facts', see Mayer-Abich (1980).

14 Communication of 28 November 1990.

15 Other sectors, especially transport, are mentioned in the Commission's documents. However it is stressed on various occasions that energy is not only the main cause of global warming but that energy policy is also the key to arresting global warming (see, for example, Communication of 28 November 1990, p. 2).

16 Environmental policy was not included in the Treaty of Rome of 1957 establishing the European Community. Such policy started with the adoption of the first environmental action programme in 1973 (OJ C 112, 20 December 1973) and was then included in the Single European Act (SEA) adopted in 1987 which amends the original Treaty. On EC environmental policy see Haigh 1992, Krämer 1990, Liberatore 1991, Rehbinder and Stewart 1985.

17 On EC energy policy, see Daintith and Hancher 1986.

18 The intention to provide international leadership in the environmental field was repeatedly stated, although not successfully implemented, in EC documents and by the former EC environment Commissioner C. Ripa di Meana in the perspective of UNCED. Emphasis on external relations also characterizes the Maastricht Treaty.

19 See the table published by the Intergovernmental Panel on Climate Change, Working Group 1 on 'Characteristics of the greenhouse gases' (IPCC WG1 1990b: p. 13):

Characteristics of the greenhouse gases

Gas	Major contributor?	Long lifetime?	Source known?
Carbon dioxide	Yes	Yes	Yes
Methane	Yes	No	Semi-quantitatively
Nitrous oxide	Not at present	Yes	Qualitatively
CFCs	Yes	Yes	Yes
HCFCs etc.	Not at present	Mainly no	Yes
Ozone	Possibly	No	Qualitatively

Such table is referred to by the EC Commission which then concludes that, 'CO_2 and CFCs are obviously the gases most relevant for immediate action' (Communication of November 1990: p. 9).

20 Such strategy means that precautionary measures should be taken, but that those measures that are also economically beneficial should be selected. Interestingly enough, the Bush Administration also advocated a 'no-regret' strategy, but with the more restricted meaning of not deciding any measure (such as stabilization targets like the one decided by the EC) that could involve high costs: costs that could be regretted if global warming should prove not to be a serious threat.

21 See OJ C 101, 24 April 1989.

22 See OJ L 222, 12 August 1988.

23 EC Commission COM (90) 365 final.

24 Mentioned in the Commission's Communication of 31 May 1991.

25 See OJ L 359, 8 December 1989.

26 This was more than one-third of the 115 million ECU allocated to the overall environment programme. Previously, only a quarter of the overall figure went to EPOCH.
27 Research on atmospheric chemistry and on other areas relevant to the climate change issue had already been started.
28 See EC Commission 1991, p. 16.
29 According to article 3b of the Maastricht Treaty, 'In areas which do not fall within its exclusive competence, the Community shall take action, in accordance with the *principle of subsidiarity*, only if and in so far as the objective of the proposed action cannot be sufficiently achieved by the member states and can, therefore, by reason of the *scale* or *effects* of the proposed action, be better achieved by the Community' (emphasis added).

BIBLIOGRAPHY

Capros, P., Karadeloglu, P., Manztos, L. and Mentzas, G. (1991) *Impact of Energy and Carbon Tax on CO₂ Emissions*, Report for DG XII of the EC Commission, Joule Programme.

Coherence, D. (1991) *Cost-effectiveness Analysis of CO₂ Reduction Options*, Report for DG XII of the EC Commission, Joule Programme.

CRU and ERL (1990) *The Development of a Framework for the Evaluation of Policy Options to Deal with The Greenhouse Effect*, Working Paper 1: *ESCAPE framework definition and work programme*, London: CRU/ERL.

Daintith, T. and Hancher, L. (1986) *Energy Strategy in Europe: The Legal Framework*, Berlin: De Gruyter.

EC Commission (1980) *Official Journal of the EC*, OJ L 12, 17 January.

—— (1991) *The European Community Crossroads in Space*, EUR 14010, Luxembourg: EC.

EI (Environment Institute) (1990) *Annual Report 90*, EUR 13806 EN, Joint Research Centre.

—— (1991) '1992–1994 activities', Internal document, Joint Research Centre.

Energy in Europe (1990) *Energy for a New Century: The European Perspective*, Special issue, DG XVII of the EC Commission.

ESA (1988) 'Objectives and strategy for the earth observation programme of the European Space Agency', Directorate of Earth Observation and Microgravity, Internal document, Paris.

—— (1991) *ESA Bulletin*, ERS-1 Special issue no. 65.

Fantechi, R. and Margaris, N. (1986) *Desertification in Europe*, Dordrecht and Boston: Reidel.

—— and Ghazi, A. (1989) *CO₂ and Other Greenhouse Gases: Climatic and Associated Impacts*, Dordrecht and Boston: Reidel.

Haigh, N. (1992) 'The European Community and international environmental policy', in A. Hurrel and B. Kingsbury (eds), *The International Politics of the Environment*, Oxford: Clarendon Press.

Krämer, L. (1990) *EEC Treaty and Environmental Protection*, London: Sweet & Maxwell.

IPCC WG1 (1990a) *Scientific Assessment of Climate Change*, Report of Working Group 1 to the Intergovernmental Panel on Climate Change, World Meteorological Organization.

—— (1990b) *Policymakers' Summary of the Report of Working Group 1 to the Intergovernmental Panel on Climate Change*, United Nations Environment Programme.

Liberatore, A. (1989) *EC Environmental Research and EC Environmental Policy. A Study in the Utilization of Knowledge for Regulatory Purposes*, EUI working paper, Florence.

—— (1991) 'Problems of transnational policy-making: environmental policy in the European Community', *European Journal of Political Research* 19: 623–42.

Mayer-Abich, K. (1980) 'Chalk on the white wall? On the transformation of climatological facts into political facts', in J. Ausubel and A. Biswas (eds), *Climatic Constraints and Human Activities*, Oxford and New York: Pergamon Press.

Mors, M. (1991) *The Economics of Policies to Stabilize or Reduce Greenhouse Gas Emissions: The Case of CO_2*, Brussels: DG II of the EC Commission.

Rehbinder, E. and Stewart, R. (1985) *Integration through Law*, vol. 2: *Environmental Protection Policy*, Berlin and New York: De Gruyer.

10

GLOBAL SOCIOLOGY AND GLOBAL ENVIRONMENTAL CHANGE

Leslie Sklair

INTRODUCTION

There are many possible 'global sociologies' and there are already in existence a few actual attempts to construct them. I shall restrict myself here to a summary of the version of global sociology that I have been attempting to construct,[1] and to sketching out a framework within which this global sociology might be applied to conceptualizing and researching issues of global environmental change.

One of the distinctive features of this global sociology is that it is self-consciously materialist (and thus anti state centrist) in two senses. First, it takes sides on the class versus state debate on the source and exercise of power, in so far as it hypothesizes the primacy of economic power over political and/or cultural power. Second, it introduces the concept of *transnational practices* to cope theoretically and empirically with questions that the conventional state-centrist versions of inter-national society or the system of nation-states (which often passes for a 'global' approach) cannot deal with adequately or at all. The 'global' in global sociology, therefore, involves more than the relations between states and state actors.

The next section will outline the conceptual framework of the sociology of the global system and discuss some parallel concepts for linking global sociology to questions of global environmental change. The following section will outline a research agenda for a global sociology of global environmental change. The chapter concludes with a double question: is there a contradiction between 'capitalist development' and 'global survival'; and, can the global capitalist system resolve it?

CONCEPTUAL FRAMEWORK

The aim of this programmatic essay is to establish the conceptual viability of a *global sociology of environmental change*, based on a specific conception of the *global system*.

The 'natural' approach to the whole world is state-centrist in a dual sense, in so far as it emphasizes the role of the state and cognitively privileges the system of nation-states. While not ignoring the nation-state, the approach proposed here offers in addition a conception of the global system based on *transnational practices*. Transnational practices are practices that cross state boundaries but do not necessarily originate at the level of the state. They are analytically distinguished in three spheres: economic, political and cultural-ideological. In the concrete conditions of the world as it is, a world largely structured by global capitalism in its various guises, each of these practices is typically, but not exclusively, characterized by a major institution. The transnational corporation (TNC) is the major locus of transnational economic practices; what I term the transnational capitalist class (TCC) is the major locus of transnational political practices; and the major locus of transnational cultural-ideological practices is to be found in the culture-ideology of consumerism (C-I of C). While there is general agreement that the transnational corporation is a key institution in the world today, the transnational capitalist class and the culture-ideology of consumerism are not established concepts and thus require justification.

The *transnational capitalist class* consists of the members of economically dominant groups and those who provide specialist services for them. The various parts of the TCC see their own interests and/or the interests of the countries of which they are citizens, as best served by an identification with the interests of the global capitalist system, as exemplified by the transnational corporations that dominate the global economy. The capitalist global system is fundamentally unstable in that the property, the political power and the culture-ideology of the transnational capitalist class are under constant challenge. How the transnational corporations mobilized through the transnational capitalist class, and act to perpetuate their interests by promoting the culture-ideology of consumerism, is the object of the many struggles for the global system. Who will win and who will lose these struggles is not a foregone conclusion. This is the significance of the TCC.

Not all culture is ideological, even in capitalist societies. The elision of *culture-ideology* is intended to signal a particular set of practices in contemporary global capitalism – namely, the institutionalization of consumerism. Culture always has ideological functions for consumerism in the capitalist global system, so all cultural transnational practices in this sphere are at the same time ideological transnational practices, and thus cultural-ideological. Cultural-ideological transnational practices and, in particular the culture-ideology of consumerism in the global system, are conceptual tools in a theory of the global system. The function of these tools is best illustrated through the proposition that global capitalism resists cultural challenges. Those cultural practices that cannot be incorporated into the culture-ideology of consumerism (that is, commercialized) become oppositional counter-hegemonic forces, to be rendered safe by marginalization, and if that fails, destroyed physically.

Ordinary so-called 'counter-cultures' are regularly incorporated and commercialized and pose no threat; indeed, through the process of differentiation (both real and illusory variety and choice) they are a source of great strength to the global capitalist system. The only counter-cultures that present potential threats to global capitalism at present, now that Stalinist communism is thoroughly discredited, are Islamic fundamentalism and the 'green' or environmentalist movement. The case of Islamic fundamentalism is beyond the scope of this chapter. As I shall argue below, there is growing evidence that central parts of the 'green movement' are in the process of being incorporated, and those that refuse incorporation are being marginalized. The *Green Consumer Guide* has replaced *Small is Beautiful*! Nevertheless, the global capitalist system has a challenging though immensely less powerful rival in (what I shall loosely call) the 'global environmentalist system'.[2]

The global capitalist system exists in order to ensure the conditions for the continued accumulation of capital and its legally guaranteed conversion into private wealth. This it does globally through the economic institution of the TNC, the political institution of the transnational capitalist class, and the culture-ideology of consumerism. How does the global environmentalist system work? Its characteristic institutional forms are what can be termed *transnational environmentalist organizations*, *transnational environmentalist elites* and the *culture-ideology of environmentalism*.

These labels are quite artificial. They are derived by analogy from the framework of my version of global sociology. To provide some evidence for the existence of these institutional forms of the global environmentalist system and to work out to what extent they are global forces in comparison with and/or in opposition to the global capitalist system, are the tasks of this chapter.

Assuming that the global environmentalist system exists, it is important to be clear why it exists. Briefly, my argument is that it exists to resolve eight critical environmental issues, globally and locally. These issues are:

- quality of the atmosphere;
- water quality;
- loss of soil productivity;

- loss of genetic diversity;
- deforestation;
- toxic contamination;
- hazardous materials;
- depletion of indigenous and dependence on imported resources.[3]

While fully recognizing that the global capitalist system is not the only source of environmental crises and in some cases not even the main source (for example, where state agencies are responsible), a global sociology of environmental change will have to tackle the questions of the impact of the global capitalist system (TNCs, TCC, and consumerism), on these eight critical issues. In order to answer these questions we must start the journey from relatively neat conceptual frameworks to a problematic and open-ended research agenda.

RESEARCH AGENDA

Economic transnational practices

Economic transnational practices are economic practices that transcend state boundaries. One important consequence of the expansion of the global capitalist system has been that individual economic actors (like workers and entrepreneurs) and collective economic actors (like trade unions and TNCs) have become much more conscious of the transnationality of their practices and have striven to extend their global influence. The volume of economic transnational practices has increased phenomenally since the 1950s, as evidenced by the tremendous growth of foreign trade. According to World Bank data global exports and imports rose from under US$100 billion in 1965, to US$3,300 billion (US$3.3 trillion) in 1981 to about US$6 trillion in 1989 (*World Development Report*, various issues). Even allowing for inflation, this is a huge increase. People all over the world can now distinguish many consumer goods in terms of their origins and the status-conferring advantages that some origins have over others. A sociology of imports, exports and global brands raises interesting possibilities.

The largest TNCs have assets and annual sales far in excess of the gross national products of most countries in the world. In 1990, about 60 countries (excluding Eastern Europe and all countries with under 1 million people) had GNPs of less than US$10 billion (World Bank 1992: Table 3). *Fortune* magazine's 'Global 500 Industrial Corporations' (July 1992) shows 135 TNCs with annual sales in 1990/91 in excess of US$10 billion. Thus, such well-known companies as Ford, General Motors, Shell, Toyota, Volkswagen, Nestlé, Sony, Pepsico,

Coca-Cola, Kodak, Xerox (and many others most of us have never heard of) have more economic power at their disposal than the majority of the countries in the world (see also, United National 1988). These figures prove little in themselves, they simply indicate the *gigantism* of TNCs relative to most countries and suggest, intuitively that their environmental impacts must be substantial.

Not only have TNCs grown enormously in size in recent decades but their 'global reach' has expanded dramatically. Many of the biggest TNCs now earn more than half of their revenues from foreign sales, and more and more TNCs are doing more and more of their manufacturing in foreign countries, including many Third World or newly industrializing countries (NICs).

All TNCs have environmental impacts, but those in some sectors have been identified as having many more and more serious effects than others (see Leonard 1988). Though global capitalism cannot be held solely responsible for all the environmental crises that threaten the survival of life on earth today, if we are to research these crises sociologically then we must at least consider the role of the global capitalist system in producing, reinforcing and resolving them. One of the first items on the research agenda of a global sociology of environmental change would be to identify the environmental effects (ranging from minimal, significant, substantial to catastrophic) of TNCs (or the sectors TNCs are typically involved in). A preliminary list would clearly include those sectors identified in Table 10.1. The contents of the 'environmental impact' boxes are intended to be suggestive,

Table 10.1 TNC impacts on environmental crises

	AT	WQ	SP	GD	DF	TC	HM	RES
OIL	mi–si	si	si	su–c	mi	si	si	su–c
CHEM	su–c	su–c	si–su	su–c	mi	su–c	su–c	si
MIN	si–su	si	si	mi	mi–si	si	su–c	su–c
EL	mi–si	si–su	mi	mi	mi	su–c	su–c	mi
LOG	mi–si	mi	si	su	su–c	mi	mi	su–c
FOOD	mi	si–su	su–c	si	si–su	si	su–c	su
AUT	su–c	mi	si	mi	mi–si	si	si	si

Notes: Sectors are: oil, chemicals, mining, electronics, logging, food processing and automobiles.
Crises are: quality of the atmosphere (AT), water quality (WQ), loss of soil productivity (SP), loss of genetic diversity (GD), deforestation (DF), toxic contamination (TC), hazardous materials (HM), depletion of indigenous/dependence on imported resources (RES).
Impacts are: mi = minimal impact; si = significant; su = substantial; c = catastrophic.

not definitive, in that they represent hypotheses to be researched rather than the findings of research already conducted, though there is already a solid body of findings in the literature on the environmental impacts of TNCs (see, for example, Allen 1992, Carson 1962, Ellis 1989, Kloppenberg 1988, Leonard 1988, Pearson 1987, Sklair 1994).

No doubt more sectors and environmental crises could be added to the list in Table 10.1 and specific TNCs could be identified as causing and/or solving specific crises (see Willums and Goluke 1992), and a more precise quantitative assessment of impacts could be developed. We could also try to distinguish between local and global environmental impacts, though this is not always possible. My intention here is to try to provide one possible framework.

Much of this research has, of course, already been embarked upon (see works cited above and many others for numerous examples). What distinguishes a global sociology of the environment from a systematic exposé of environmental hazard is the attempt to contstruct an explanation of how the global capitalist system comes to engage in what, on the surface, appears to be entirely irrational behaviour in destroying its sphere of operations – namely, the global environment.

One way to attack this problem is to draw out the relationships between the characteristic institutions of global capitalism, the TNCs, and those of the global environmentalist system, the transnational environmentalist organizations (TEOs). These relationships straddle the spectrum from direct and indirect TNC sponsorship and support of TEOs to downright hostility and occasional violence between the parties and their allies. A first approximation of TEOs would certainly include the United Nations Environment Programme, the World Conservation Union (formerly International Union for the Conservation of Nature), WorldWide Fund for Nature, Friends of the Earth, Greenpeace and the International Organization of Consumers Unions.[4] To this list must be added the myriad of global environmentalist organizations that mushroomed around the Brundtland Report and the 1992 UN Conference on Environment and Development in Rio de Janeiro and its aftermath. There is evidence that at least some TEOs are already collaborating with, if they have not actually been co-opted by, the TNCs.[5] I predict that those who refuse to be co-opted will be marginalized. This is clearly an important issue for researchers on global environmentalist movements (see Viola and Vieira 1991).

At the national and local levels this process appears to be accelerating (see Elkington and Burke 1989). In an informative article, *Advertising Age*, the premier US advertising magazine, recently listed the tie-ins between the hundred leading national advertisers and environmental groups, detailing cash donations and promotional programmes. This makes very interesting reading (see *Advertising Age* 1991). The extent to which local, grass-roots environmentalist groups can resist these trends is clearly a very fruitful topic for research.[6]

This leads on directly to a consideration of the transnational capitalist class (TCC) and what I have ventured to label the transnational environmentalist elite (TEE).

Political transnational practices

The transnational capitalist class is not made up of capitalists in the traditional Marxist sense. Direct ownership or control of the means of production is not the exclusive criterion for serving the global interests of capital. The political practices of the transnational capitalist class may be analysed in terms of two issues. First, how does it operate to change the nature of the political struggle between capital and labour? This can be measured in terms of its domestic and transnational political organization, direct and indirect TNC interference in host country politics, and the extent to which these constrain and are constrained by the local and/or transnational labour movement. Second, how does the transnational capitalist class aim to downgrade certain domestic practices by comparison with new and more lucrative transnational practices and to create a global perspective? This can be measured by the local and in some cases international brain drain from domestic to transnational enterprises, mainly but not exclusively TNCs. The people who make up this brain drain play a crucial role for the transnational capitalist class in persuading co-nationals that their interests are identical with, or at least best served by, those of the TNCs.

We can identify at least four separate though more or less integrated factions of the transnational capitalist class, globally and locally. These are:

- TNC executives and their local affiliates;
- globalizing state bureaucrats;
- capitalist-inspired politicians and professionals;
- consumerist elites (merchants, media).

This class sees its mission as organizing the conditions under which its global and local interests (which usually, but do not always, coincide) can be furthered. The concept of the TCC implies that there is one central *transnational* capitalist class that makes system-wide decisions, and that it connects with the TCC in each locality, region and country.

Although comprador classes in one form or another have existed for centuries, the transnational capitalist class is a relatively new

211

phenomenon. The basic difference between the two is that whereas compradors are entirely beholden to the TNCs and foreign interests, the 'local branch' of the TCC can develop into a class that can, under certain circumstances, begin to dictate its own terms to the TNCs and foreign interests. The logical extension of this argument is that some form of interdependence is possible, where TCCs in Third World countries could carve out niches for themselves in the crevices, sometimes even the chasms, that the hegemonic TNCs leave unattended.

The four groups of people who would constitute a 'transnational environmentalist elite' parallel the four factions of the transnational capitalist class. They are:

- transnational environmental organization executives and their local affiliates;
- globo-localizing 'green' bureaucrats;
- 'green' politicians and professionals;
- 'green' media and merchants.

The analysis of these four groups must be carried out in the context of the conjuncture of economic and environmental crises that have been afflicting most countries of the world in recent decades.[7]

Some sections of the transnational capitalist class do not see the environmental 'crises' as critical, because of their belief in the inherent creative capacity of global capitalism to solve more or less any problem. This is very well illustrated in the controversy sparked off by the Club of Rome report on the 'limits to growth' of 1972.[8] For many, 'the population problem' lies at the root of the so-called environmental 'crises' of our age.[9] For the transnational environmentalist elite (TEE) the answer is, naturally, not so simple, and just as each fraction of the TCC serves the interests of the capitalist global project in its own particular fashion, each fraction of the global environmentalist elite serves the interests of global environmental protection and enhancement in its own fashion.

The TEO executives tend to be from backgrounds in international organizations (particularly agencies of the UN), and increasingly in TNCs and other capitalist global organizations. The hypothesis that 'TNCs are providing an increasing proportion of the numbers and strategies of TEO executives' would certainly be worth researching. The local affiliates of the TEOs tend to be from public service (particularly the research and educational sector) or radical political backgrounds.[10]

Second, 'globo-localizing green bureaucrats' play a crucial role in the struggle that ensues when the interests of global capitalism and the global environment are in conflict. This group consists largely of those who have an apparent official duty and moral commitment to protect and enhance the environment against the onslaughts of those who would damage it. Again, our research agenda requires careful study of these groups and the circumstances in which they work.

Third, we must consider the green politicians (members of 'green parties', and the increasing numbers of non-green party politicians who for one reason or another have jumped on the 'green band-wagon'); and green professionals (lawyers, academics, journalists, ethical investment analysts, consultants, etc. who are dedicated to spreading the environmentalist message).

The fourth part of the transnational environmentalist elite consists of the 'green' media and merchants, by which I mean those selling and those involved in media promotion of environmentalism. The undisputed fact that the 'green consumer' is now very big business (see, for example, Czinkota and Ronkainen 1991, Elkington and Hailes 1988, Wells and Jetter 1991) seems to me to be a crucial component of any global sociology of the environment. The potential for alliances and conflict between the merchants and media of global capitalism and the 'green' merchants and media of the global environmentalist elite, and the ways in which the culture-ideology of consumerism deals with environmental issues, are research-rich areas of inquiry.

These, then, are the fractions of the TEE and, if my analysis is correct, the traditional 'back to nature' and 'conservation' movements that previously spoke for and nurtured the environment are being transformed as a consequence of the relentless insertion of the whole world into the global capitalist system. It remains an open question whether and to what extent the global environmentalist elite can mount an effective challenge to protect and enhance environmentalist interests in the face of the unrestrained expansionism and resource profligacy that is inherent to the global capitalist project.

One set of issues that has been on the research agenda for some time are the relations between environmental regulatory bodies and business, both globally and at the level of the state. Globally, the history of attempts to regulate the environmental impacts of business through international and transnational organizations, largely but not entirely parallel with the United Nations system since the 1950s, are well documented (see, for example, McCormick 1989). Two important recent developments are the 'Benchmark Corporate Environment Survey' of the UN and the Montreal Agreement on CFCs. The 'Benchmark Survey' was a project of the UN Centre on Transnational Corporations to discover what environmental policies TNCs currently had and what TNCs were doing to improve their environmental performance. Data from 163 TNCs with sales of over US$1 billion have been collected (United Nations 1991).[11]

Through the Montreal Protocol of 1987 on the protection of the ozone layer, in the words of the US negotiator: 'the signatory countries sounded the death knell for an important part of the international chemical industry, with implications for billions of dollars of investment and hundreds of thousands of jobs in related sectors' (Benedick 1991: 1). If this is indeed the case, then this would count as a strong counter-example against my thesis that the TNCs of the global capitalist system are hegemonic over the so-called 'inter-national community'. The story of CFCs and other ozone-depleting chemicals and attempts to substitute them, however, is far from over. It is, nevertheless, a crucial case.

At the level of the state, Schrecker (1990) argues that the phenomenon of poacher turned gamekeeper (my phrase, not his) is becoming common in government appointments to environmental agencies. (Schrecker's example is the appointment of an Imperial Oil/Exxon executive as adviser to Canada's Minister of the Environment). However, it is clear from the evidence of the 1980s that governments, even anti-regulatory right-wing ones like the Reagan and Thatcher Administrations, can no longer ignore environmental violations. For example, while the Reagan Administration was pulling the teeth of the Environmental Protection Agency in the USA, at the same time it permitted the establishment of a powerful Environmental Crimes Unit in the Department of Justice. Schrecker speculates on a potential 'environmental corporatism of the future' and describes the Canadian National Task Force on Environment and Economy (a response to Brundtland) in terms that are worth quoting. It, he says,

> comprised the federal Environment Minister and a number of his provincial counterparts, one academic, two representatives of non-governmental environmental organizations, the head of the Ontario Waste Management Corporation, and business representation including senior executives with some of the worst polluters (and most aggressive opponents of environmental regulation) in Canada. The Task Force completely excluded representatives of ogranized labor, thereby indicating the government's perception of their irrelevance.
>
> (Schrecker 1990: 184)

This is a good approximation of the local wing of what I mean by the 'global environmentalist elite', though it is not always necessary to exclude labour movement representatives. Job blackmail is a very common ploy when business is threatened with serious enforcement of environmental laws. In such cases, an alliance of 'Workers, corporate managers and local businessmen tend, with some exceptions, to line up on the same side of such issues. The other side . . . are often from outside the immediate area, and their immediate livelihoods are conspicuously *not* at stake' (Schrecker 1990: 186). Environmental politics is a distinctive form of class politics where the global and the local frequently intersect.[12]

The issue of labour's response to environmental problems is taken up by Siegmann (1985) in a comparison of the USA and West Germany. He puts forward several explanations of the changing attitudes and practices of the two labour movements, and documents some softening in the traditional opposition of labour to environmental regulation. He argues that the forced choice between saving jobs and saving the environment (particularly in the energy sector) began to dissolve in the 1970s (the United Automobile Workers Black Lake Conference in 1976 and the anti-nuclear campaign being two key reasons for this change). Structural factors, like the rise of a separate 'green party' in Germany and the secular decline in smokestack industries in the USA, are also relevant.[13]

A three-country study of the public and elites in the US, the UK and West Germany adds some useful information and ideas on how to tackle the political alliances likely to result from the impact of environmentalism in its various forms (Milbrath 1984). Three key beliefs distinguish the environmental vanguard from the rearguard. These are based on choices on the statements:

- 'I perceive condition of environment to be no problem vs large problem';
- 'necessity for scientific-technological change vs necessity for basic social change';
- 'no limits vs limits to growth'.

Milbraith identifies eight groups as follows:

- *rearguard* (about 20 per cent of the sample): small problem, better technology, no limits to growth;
- *establishment* (10 per cent): small, better technology, limits;
- *weak establishment followers* (10 per cent): small, basic change, no limits;
- *undecided middle* (18 per cent): small, basic change, limits;
- *nature conservationist establishment followers* (5 per cent): large problem, better technology, no limits;
- *nature conservationists* (7 per cent): large, better technology, limits;
- *young, lower-class environmentalist sympathizers* (6 per cent): large, basic change, no limits;
- *vanguard* (24 per cent): large, basic change, limits.

215

Group profiles reveal, significantly, that the rearguard in the US 'is made up of many powerful people' (Milbraith 1984: 55), and that the vanguard has a sizeable proportion of *all but business leaders*, including many service sector personnel and women, in all three countries. Milbraith argues that as the service sector grows, the vanguard will grow. Some may find this research a little naïve, politically, but as most of the interviews were done at least ten years ago it would certainly be interesting to replicate the study today to ascertain whether or not the 1980s actually did see massive and irreversible changes in attitudes and practices concerning the environment, as some activists argue.[14]

In both the national and the global context, two Brazilian researchers, Viola and Vieira (1991), have developed a useful and coherent framework for studying the growth of environmental movements and their elites. They distinguish:

1 *the new social movements approach*, identified with the socialist-anarchist perspective;
2 *the historical movements approach*, based on the unsustainable character of modern industrial society; and
3 *the interest groups approach*, in which environmentalism is seen as a special case of lobbying.

Viola and Vieira favour the 'historical movements approach' and they develop a concept of *environmental movement*. The success of this movement is explained in terms of arguments about population pressure, depletion of resources, pollution and the effects of consumerism. They identify networks of 'greenish' businessmen and managers, scientific groups and institutions, bodies of public administrators, social-environmentalists, and leaders of intergovernmental agencies. Parallel to the national movements, the 'global environmentalist organizations and elites' that I have been discussing in this chapter are similar to Viola and Vieira's 'global transnational environmental movement'.

These, then, are some ongoing contributions providing empirical support and/or conceptual clarification for the central issues involved in researching the economic and political spheres of the global environmentalist system. Naturally, none of them is couched in terms of the sociology of the global system elaborated in this chapter, but they all have something to add to it.

Cultural-ideological transnational practices

The 1970s and 1980s witnessed an unprecedented increase in the scale and scope of the electronic media of communication, as well as genuine innovations in their nature. Technological advances, international competition, and consequent relative price reductions in producer and consumer electronics have led the US, European, and Japanese TNCs, that for the most part control the world's electronic media, to develop global strategies for the establishment and development of their various hegemonic practices that would have been technically impossible, and in some cases even unthinkable a few decades ago (see McPhail 1987). This gives the potential for distribution of messages on a scale never before achieved.

The technical fact that a greater variety of messages may be broadcast on a vastly greater scale than ever before does not alter the political fact that the central messages are still, and more powerfully, those of the capitalist global system. McLuhan's famous 'the medium is the message' is true to the extent that transnational corporations increasingly control the media to propagate *their* message, as McLuhan himself occasionally acknowledged. An excellent indicator of this is the phenomenal increase in commercial sponsorship of what once used to be considered purely cultural events, like operas, museum exhibitions, and sports. The commercialization of the supposedly amateur Olympic Games by some of the world's largest TNCs is a paradigm case.

Bagdikian (1989) characterizes those who control this system as 'the Lords of the Global Village'. They purvey their product (a relatively undifferentiated mass of news, information, ideas, entertainment, and popular culture) to a rapidly expanding public, eventually the whole world. National boundaries are growing increasingly meaningless as the main actors (five groups in 1989) strive for total control in the production and marketing of what we can call the cultural-ideological goods of the global capitalist system. Their goal is to create a 'buying mood' for the benefit of the global troika of media, advertising and consumer goods manufacturers. By age 16, the average North American youth has been exposed to more than 300,000 commercials. 'Nothing in human experience has prepared men, women and children for the modern television techniques of fixing human attention and creating the uncritical mood required to sell goods, many of which are marginal at best to human needs' (Bagdikian 1989: 819).

The mass media perform many functions for global capitalism. They speed up the circulation of material goods through advertising, which reduces the time between production and consumption. They begin to inculcate the dominant ideology from an early age, in the words of Esteinou Madrid, 'creating the political/cultural demand for the survival of capitalism' (in Atwood and McAnany 1986: 119). The systematic blurring of the lines between information, entertainment, and promotion of products lies at the heart of this practice. This has

217

not in itself created a culture-ideology of consumerism, which has been in place for at least the last century – perhaps longer in the First World and among comprador classes elsewhere. What it has created is a reformulation of consumerism that transforms all the public mass media and their contents into opportunities to sell ideas, values, products – in short, a consumerist world view.

What, then, is the alternative culture-ideology of environmentalism, and in what ways does it challenge the dominant culture-ideology of consumerism?

There appears to be a growing consensus among students of environmentalism that there is no single environmentalist ideology (see, in general, Dobson 1990, Milbrath 1989, O'Riordan 1981). That there are shades of green is widely accepted. O'Riordan (1991: 6), for example, argues that environmentalism has evolved through dry-, shallow- and deep-green phases, oriented to the problem of sustainable development. 'Dry greens' believe in the manipulation of the marketplace through 'benign self-regulation'; 'shallow-greens' criticize this 'reinforcing of the pernicious status quo' and focus on community-based reform, eco-auditing and environmentally benign consumerism; and 'deep greens' reject, by implication, what I have been calling the culture-ideology of consumerism and the whole global capitalist project.

Though it is tricky to try to identify particular individuals as 'dry', 'shallow' or 'deep', following O'Riordan one may suggest that much of 'green business' holds to the dry variety. For example, Milbrath's 'nature conservationist establishment followers' (those who accept that there is a large environmental problem but consider that it can be solved technologically without basic changes to social structures and that there are no limits to growth) might qualify as 'dry greens' (Milbrath 1984: ch. 3). Straight 'nature conservationists' who accept that there are limits to growth might also hold this ideology. Similarly, Schrecker (1990) reports an increasing number of industrialists who accept the need for environmental action, as long as they do not have to pay for it themselves. Again, this position is quite compatible with 'dry greenism'.

'Shallow green' ideology seems most consistent with what has come to be known as 'free market environmentalism' (see, for example, Anderson and Leal 1991, Beckerman 1975). This basically involves counting the true costs of pollution and making the polluters pay a market price for their 'use' of the environment. The most sophisticated version of this is to be found in the 'Blueprints' of Pearce and his colleagues (see Pearce *et al.* 1991). Here, the instruments for ensuring environmental protection and enhancement are a combination of 'command and control' mechanisms and 'market-based instruments'

218

to discourage polluters and encourage the introduction of non-polluting technologies through the price system. The key to future environmental success, therefore, is to provide *incentives not to pollute*.

It is tempting to hypothesize that 'shallow green' is actually a more consistent ideology for the core TNCs and countries of global capitalism than its 'dry green' alternative. As populations become richer and what is prioritized in 'quality of life' indicators changes from the satisfaction of basic needs to more 'advanced' patterns of consumption, then we might expect that an environmental dimension – 'green consumerism' – will be incorporated into the culture-ideology of consumerism. This is, indeed, what is already happening, as I suggested above. Therefore, if the culture-ideologies of consumerism and ('dry' and 'shallow') environmentalism are quite consistent belief-systems, then on the level of culture-ideology we should not be surprised to discover that the TNCs and the transnational environmentalist organizations can work well together most of the time, and that strategic alliances are being formed between the transnational capitalist class and the global environmental elite. I would expect the UNCED 'Earth Summit' in Brazil in June 1992 and its various aftermaths to provide further evidence for this view.[15]

Does 'deep green' ideology offer a real alternative to global capitalism and the culture-ideology of consumerism? Dobson (1990) usefully distinguishes 'environmentalism' from 'ecologism', and he traces 'Deep Ecology' from Arne Naess's seminal 1973 paper where the distinction between 'shallow' instrumental concern with pollution and resource depletion and 'deep' intrinsic concern with ecological principles was first drawn systematically. Dobson argues that the debate has taken two turns since 1973: namely, the development of a theory of the intrinsic value of the environment and, subsequently, a critique of this in terms of the need for a change of consciousness with respect to our relationship to the 'natural' world. Deep Ecologists are radically non-anthropocentric in the sense that human beings are no longer seen as the centre of the universe but are replaced by something else, in one version growing in appeal: 'giving way to gaia' (see O'Riordan 1991: 6 and *passim*).

There is little doubt that 'deep green' or 'ecologism' does represent a radical rejection of global capitalism and the culture-ideology of consumerism, but its viability as a practical alternative to these is at present inhibited by its lack of a programme that would appeal to large numbers of people, particularly those who are presently without ready access to the consumer goods enjoyed by the majority of people in the countries of the First World.[16] This is why, in my view, it is so important to begin to research the environmental implications of the culture-ideology of consumerism in the Third and (what was) the Second World.[17]

CONCLUSION

Is there a contradiction between 'capitalist development' and 'global survival'; and can the global capitalist system resolve it?

Almost a decade ago, Redclift put forward the then rather unfashionable argument that 'it is the poor [of the South] who are the losers in the process of environmental depredation' (1984: 2). Viola and Vieira (in 1991) are representative of a growing body of opinion that sees the environmental crisis exposing contradictions in the inner logic of industrial (if not capitalist) societies. The two poles of the contradiction are 'pollution by poverty' and 'overdevelopment'. Thus, 'it seems insufficient to formulate development policies taking into account the environmental dimension if no effective measures are provided to take equity as the distributional principle and the satisfaction of basic human needs as the main societal objective' (Viola and Vieira 1991: 3). Their emphasis on the 'planetarization of western materialist culture' calls for demand- and supply-side analyses.

However, it is exactly at this point that the key contradiction between capitalist development and environmental survival emerges. To put the issue bluntly, most researchers now seem convinced that any attempt to raise Third and Second World consumption standards to the level of that of the First World, or even to maintain present First World levels, is insupportable on environmental and resource grounds. Goulet (in Engel and Engel 1990) documents the increasing volume of anti-development critique of the 1970s and 1980s. Indeed, the recent Brundtland-inspired campaign for the concept of 'sustainable development' can be interpreted, at least in part, as an attempt to shore up some idea of 'development' against this critical tide of cultural survivalism, ecofeminism, and ecophilosophy (see De La Court 1990, Engel and Engel 1990).

The hypothesis that there is a contradiction between capitalist development and global survival appears, therefore, to have prima facie plausibility. The transformation of the 'green vanguard' of global capitalism from dry to shallow environmentalism (sustainable development) signals the fact that this message is getting through to the transnational capitalist class and that serious attention is beginning to be paid to programmes that will disconfirm the hypothesis – what Robins and Trisoglio (1992: 165–71) call 'the partial greening of business' – as evidenced by such initiatives as the International

Chamber of Commerce's World Industry Council for the Environment' (see Willums and Goluke 1992), increasing interest in systematic recycling (Vandermerwe and Oliff 1991) and the UNCTC project referred to above.

Clearly, the dominant forces in the global capitalist system have no option but to believe and act as if this contradiction can be resolved by a combination of economic-technological, political and culture-ideology means. Part of this must involve the ways that the global capitalist system uses the Third World to resolve the contradiction. This being so, I shall conclude with a concrete and very pertinent example to illustrate this contradiction.

Traditionally, some TNCs have used the Third World as a place to relocate the dirtiest and most hazardous production processes, as 'pollution havens'. While not all, possibly not even most, TNC factories in the Third World are of this type, and some TNC factories certainly achieve higher environmental standards than domestic factories, 'pollution havens' and their corollary 'industrial flight' clearly exist.[18] This is certainly one way that the TNCs could continue to resolve the contradiction for the countries of the First World, but a combination of international political pressures, the global nature of many environmental hazards and the unwillingness of Third World governments and their peoples to tolerate such behaviour, makes it unlikely that this could be a long-term solution for the core capitalist countries.[19]

A horrific aspect of this problem is what the UNCTC has labelled 'unregulated waste tourism'. The Basel Convention on the Control of Transboundary Movements of Hazardous Waste and Their Disposal (1989) was an attempt to regulate abuse in this field, but central to the question is the myth that everything is cheaper in the South. Labour is, but the other inputs for hazardous waste management are at world or even higher prices, so the foreign currency shortage that drives Third World governments to such desperate measures may increase rather than decrease as a result of these activities (see Slaughter 1990). Already, we can see such problems emerging in the international attempts to implement the well-intentioned 'Conventions' and 'Agenda 21' aspirations of the Rio Earth Summit. The hypothesis that it is the culture-ideology of consumerism that drives the system and, thus, provokes the contradiction between capitalist development and global survival, helps us to explain why TNCs produce and treat their wastes in such a cavalier fashion and why governments and elites in the Third World are so desperate for foreign currency, partly to finance imports of consumer goods or the means to produce them .[20]

For too long the social study of global environmental change has been focused on supply-side issues (predominantly production systems), and has paid too little attention to demand-side issues (particularly consumption patterns and the culture-ideology that transforms and sustains them). While recognizing the essentially dialectical relationship between 'supply' and 'demand' in capitalist societies, a central aim of the global sociology of global environmental change put forward here is to try to redress this imbalance.

NOTES

1 Principally in my book, *Sociology of the Global System* (1991) from which this chapter borrows. See also, Sklair (1992). For a specifically 'environmental sociology' approach see Buttel and Taylor (Chapter 11, this volume).

2 Just as there are many differing and even antagonistic parts of the global capitalist system, the global 'green' system is not all of a piece (see O'Riordan 1981). The distinction between 'environmentalism' and 'ecologism' is useful here (see Dobson 1990). I shall elaborate on this later (see pp. 218–20).

3 These are the issues that appear to me, on surveying the literature (for example, Caldwell 1985, De La Court 1990, Meadows *et al.* 1992, Pearce *et al.* 1991, Toth *et al.* 1989) to be the key issues for the system today, though no doubt other lists could be made. The only controversial omission, I think, is that of population. However, I am not alone in believing that 'the population problem' is of a different order to those I have listed.

4 The US organizations, the National Wildlife Federation, the National Audubon Society and the Sierra Club, who in 1982 were reported to have budgets of US\$37 million, US\$22 million and US\$13 million and memberships of 4.2 million, 470,000 and 311,000 respectively (data cited in Siegmann 1985: 24) would probably qualify as the largest environmentalist organizations in the world. Although they are not global as such, all have global agendas (see, for example, National Wildlife Federation 1991).

5 There is formidable big business input into the formation of the UN Commission on Sustainable Development (CSD), the major institutional result of UNCED. In particular, the role of the Business Council for Sustainable Development, which represented the TNC position in Rio (see Schmidheiny 1992), and the environmental activities of the international Chamber of Commerce (see Willums and Goluke 1992) deserve close scrutiny. CSD is destined to become a major transnational environmental organization and it will be a critical test case for my theory.

6 Allen (1992) is very instructive on this question. The magazines and newsletters of the major environmental groups also repay careful study.

7 However, the origins of this crisis can be traced back to the eighteenth century at least. For a fascinating discussion of the long-standing connections between environmentalism and colonialism, see Grove (in Angell *et al.* 1990).

8 For a comprehensive account of this see Kassiola (1990), whose discussion of the attempt of the pro-growth school to depoliticize and scientize their 'economic' value judgements is very stimulating. Meadows *et al.*

222

(1992) is an update of the original report, which now revolves around the umbrella idea of 'sustainable development' (for a useful brief critique of which see Daly 1991, especially Ch. 13; and below, pp. 220-2).

9 For example, Garrett Hardin's influential 'tragedy of the commons' critique of development policies is based on a rejection of what he calls 'the concept of the freedom to breed' (see Hardin and Baden 1977, for a review of the 'commons debate').

10 Manes, in his evocatively entitled book *Green Rage*, shows how the radical Earth First! movement grew out of 'disillusionment and exasperation with the reform environmental movement and its *idée fixe*, credibility' (1990: 66) and documents the rise of 'ecotage' (environmentally motivated sabotage).

11 I am grateful to Dr Harris Gleckman, Chief of the UNCTC Environmental Unit, for material on this ongoing project. The recent change in the status of the Centre and the reasons for it are of interest for the thesis I am developing.

12 For example, compare the glowing profile of the Chicago-based Waste Management, Inc. (an environmental services corporation with sales of US$7.4 billion) in *Fortune* magazine (13 January 1992: 46-7) with the reasons why it is 'the most heavily fined disposal firm in the country' (Lipsett 1991: 27). See also, Sklair (1994, note 17).

13 For an interesting analysis of a related class question (wittily identified as the 'greening of the rednecks') see Sagoff (1990); and, in general, Rüdig (1990).

14 In national environment surveys of adults in the US, when respondents were asked to choose between protecting the environment and protecting jobs, in 1989 52 per cent chose to protect the environment, and 36 per cent to protect jobs. By 1991, the percentages were 69 per cent and 19 per cent respectively (Environmental Opinion Study, Inc., 1991: 14). I am grateful to Jessica Mathews of World Resources Institute, Washington, DC, for access to this study. For analysis of similar survey data from the 1960s and 1970s see Vogel (1977). Allen (1992) has information on similar issues in Britain and Ireland.

15 For surveys of the preparations and outcomes of the 'Earth Summit' see *Network* (Centre for Our Common Future, Geneva). This organization is one of the several spinoffs of the work of the Brundtland Commission (for useful discussions of which compare De La Court 1990 and Angell *et al.* 1990).

16 O'Riordan excludes socialist environmentalism from his 'evolution of environmentalism'. For examples of the 'socialist environmental alternative' see Coates (1979) and Ryle (1988), but it is not at all clear how it can cope with the critique of industrialism (see Dobson 1990; ch. 5, Kassiola 1990). *Capitalism, Nature, Socialism*, a new journal of socialist ecology (Santa Cruz) reflects these tensions.

17 For a useful discussion of 'consumption and environment' see Parikh *et al.* (1991), specially prepared for but largely ignored at Rio. This is an issue that I am taking up in an international research project on 'the culture-ideology of consumerism' with colleagues in China, Brazil, South Korea and Eastern Europe.

18 The most notorious case is the Bhopal disaster (see Bhargava 1988 and Pearson 1987, particularly chs 11, 12). For a balanced comparative

analysis, see Leonard (1988). It is worth noting that the opposition to the Free Trade Agreement negotiations between the USA and Mexico has been spearheaded by environmental groups in both countries on precisely these grounds, see Sklair (1993: ch. 11). I deal in more detail with the related question of the environmental impacts of export-oriented industrialization in Sklair (1994).

19 'Just between you and me, shouldn't the World Bank be encouraging *more* migration of the dirty industries to the LDCs?' These leaked words of Lawrence Summers, chief economist and vice-president of the World Bank, hit the headlines in 1992 (cited in *The Guardian*, London, 14 February 1992: 29). As Leonard (1988) shows, economists have been arguing this for some time. A recent, thorough, World Bank-sponsored discussion is Low (1992).

20 I am, of course, aware that much of the available foreign currency is spent on weapons and other instruments of repression, but it is not difficult to connect this to the need to sustain and protect the consumption of the elites.

REFERENCES

Advertising Age (1991) 'Your guide to green groups. Where top advertisers turn for help', 28 October.

Allen, R. (1992) *Waste Not, Want Not: The Production and Dumping of Toxic Waste*, London: Earthscan.

Anderson, T. and Leal, D. (1991) *Free Market Environmentalism*, Boulder, Calif.: Westview.

Angell, D., Comer, J. and Wilkinson, M. (eds) (1990) *Sustaining Earth: Response to the Environmental Threats*, London: Macmillan.

Atwood, R. and McAnany, E. (eds) (1986) *Communication and Latin American Society. Trends in Critical Research, 1960–1985*, Madison, University of Wisconsin Press.

Bagdikian, B. (1989) 'The lords of the global village', *The Nation*, 12 June: 805–20.

Beckerman, W. (1975) *Pricing for Pollution*, Hobart Paper 66, London: Institute of Economic Affairs.

Benedick, R. (1991) *Ozone Diplomacy: New Directions in Safeguarding the Planet*, Cambridge, Mass.: Harvard University Press.

Bhargava, A. (1988) 'The Bhopal incident and Union Carbide: ramifications of an industrial accident', *Bulletin of Concerned Asian Scholars* 18(4): 2–19.

Buttel, F. and Taylor, P. (1994) 'Environmental sociology and global environmental change: a critical assessment', Chapter 11 in M. Redclift and T. Benton (eds), *Social Theory and the Global Environment*, London: Routledge.

Caldwell, L. (1985) *US Interests and the Global Environment*, Iowa, Stanley Foundation.

Carson, R. (1962) *Silent Spring*, Greenwich: Fawcett.

Coates, K. (ed.) (1979) *Socialism and the Environment*, Nottingham: Spokesman Books.

Czinkota, M. and Ronkainen, I. (1991) *Global marketing 2000: future trends and their implications. A Delphi Study*, Washington, DC: Georgetown University, for the American Marketing Asssociation.

Daly, H. (1991) *Steady-State Economics* (2nd edn), Washington: Island Press.

De La Court, T. (1990) *Beyond Brundtland: Green Development in the 1990s*, London: Zed.

Dobson, A. (1990) *Green Political Thought*, London: Unwin Hyman.

Elkington, J. and Hailes, J. (1988) *The Green Consumer Guide*, London: Gollancz.

—— and Burke, T. (1989) *The Green Capitalists: How to Make Money and Protect the Environment*, London: Gollancz.

Ellis, D. (1989) *Environments at Risk: Case Histories of Impact Assessment*, Berlin: Springer-Verlag.

Engel R. and Engel, J. (eds) (1990) *Ethics of Environment and Development: Global Challenge, International Response*, London: Belhaven Press.

Environmental Opinion Study, Inc. (1991) *'Survey'*, Washington DC, June.

Hardin, G. and Baden, J. (eds) (1977) *Managing the Commons*, San Francisco: W.H. Freeman.

Kassiola, J. (1990) *The Death of Industrial Civilization: The Limits to Economic Growth and the Repoliticization of Advanced Industrial Society*, Albany, N.Y.: SUNY press.

Kloppenberg, J. Jnr (1988) *First the Seed: The Political Economy of Plant Biotechnology, 1492–2000*, Cambridge: Cambridge University Press.

Leonard, H. (1988) *Pollution and the Struggle for the World Product*, Cambridge: Cambridge University Press.

Lipsett, B. (1991) 'Trashing the future', *Multinational Monitor*, September: 27 31.

Low, P. (ed.) (1992) *'International Trade and the Environment*, Discussion Paper 159, Washington DC: World Bank.

McCormick, J. (1989) *The Global Environmental Movement: Reclaiming Paradise*, London: Belhaven.

McPhail, T. (1987) *Electronic Colonialism*, Newling Park: Sage.

Manes, C. (1990) *Green Rage: Radical Environmentalism and the Unmaking of Civilization*, Boston: Little, Brown & Co.

Meadows, D.H., Meadows, D.L. and Randes, J. (1992) *Beyond the Limits: Global Collapse or a Sustainable Future*, London: Earthscan.

Milbrath, L. (1984) *Environmentalists: Vanguard for a New Society*, Albany, N.Y.: State University of New York Press.

—— (1989) *Envisioning a Sustainable Society*, Albany, N.Y.: State University of New York Press.

National Wildlife Federation (1991) *Trade and the Environment*, Information Pack, Washington, DC.

O'Riordan, T. (1981) *Environmentalism*, London: Pion.

—— (1991) 'The new environmentalism and sustainable development', *Science of the Total Environment* 108: 5–15.

Parikh, J. *et al.* (1991) 'Consumption patterns: the driving force of environmental stress', Indira Gandhi Institute of Development Research, Bombay (prepared for UNCED).

Pearce, D., Markandya, A. and Barbier, E. (1991) *Blueprint 2: Greening the World Economy*, London: Earthscan.

Pearson, C. (ed.) (1987) *Multinational Corporations, Environment, and the Third World: Business Matters*, Durham, N.C.: Duke University Press (A World Resources Institute Book).

Redclift, M. (1984) *Development and the Environmental Crisis*, London: Methuen.

Robins, N. and Trisoglio, A. (1992) 'Restructuring industry for sustainable development, Chapter 6 in J. Holmberg (ed.), *Policies for a Small Planet*, London: Earthscan.

Rüdig, W. (ed.) (1990) *Green Politics One*, Edinburgh University Press.

Ryle, M. (1988) *Ecology and Socialism*, London: Radius.

Sagoff, M. (1990)' "I am no Greenpeacer, but . . ." or Environmentalism, risk communication, and the lower middle class', Chapter 10 in W. Hoffman, R. Frederick and E. Petry (eds), *Business, Ethics, and the Environment: The Public Policy Debate*, Westport: Quorum Books.

Schmidheiny, P. (1992) *Changing Course*, Cambridge, Mass.: MIT Press.

Schrecker, T. (1990) 'Resisting environmental regulation: the cryptic pattern of business-government relations', in R. Paehlke and D. Torgerson (eds), *Managing Leviathan: Environmental Politics and The Administrative State*, London: Belhaven: pp. 167–99.

Siegmann, H. (1985) *The Conflicts between Labor and Environmentalism in the FRG and the US*, Berlin: WZB.

Sklair, L. (1991) *Sociology of the Global System*, London: Harvester, and Baltimore: Johns Hopkins University Press.

—— (1992) 'Globalization and Europe: theoretical frameworks and research agendas', British Sociological Association, Annual Conference, University of Kent.

—— (1993) *Assembling for Development: the Maguila Industry in Mexico and the United States* (2nd edn), University of California San Diego: Center for US-Mexican Studies.

—— (forthcoming) 'Global system, local problems: transnational corporations and environmental hazards in the Mexico–US borderlands', in H. Main and W. Williams (eds), *Environment and Housing in Third World Cities*, London: Belhaven.

Slaughter, T. Jnr (1990) 'The improbability of Third World government consent in the coming North–South international toxic waste trade, Chapter 19 in W. Hoffman, R. Frederick and E. Petry (eds), *Business, Ethics, and the Environment: The Public Policy Debate*, Westport: Quorum Books.

Toth, F., Hizsnyik, E. and Clark, W. (eds) (1989) *Scenarios of Socioeconomic Development for Studies of Global Environmental Change: A Critical Review*, Laxenburg: International Institute for Applied Systems Analysis.

United Nations (1988) *Transnational Corporations in World Development*, New York: UNCTC.

United Nations (1991) *Consolidated Executive Summaries of the Benchmark Corporate Environmental Survey*, New York: UNCTC.

Vandermerwe, S. and Oliff, M. (1991) 'Corporate challenges for an age of reconsumption', *Columbia Journal of World Business* XXVI (Fall): 6–25.

Viola, E. and Vieira, P. (1991) 'From preservationism and pollution control to sustainable development: an ideological and organizational challenge for the environmental movement in Brazil', International Seminar, Institute of Latin American Studies, London, November.

Vogel, D. (1977) 'Business and the politics of slowed growth', in D. Pirages (ed.) *The Sustainable Society: Implications for Limited Growth*, New York: Praeger.

Wells, P. and Jetter, M. (1991) *The Global Consumer: Best Buys to Help the Third World*, London: Gollancz.

Willums, J.-O. and Goluke, V. (1992) *From Ideas to Action: Business and Sustainable Development*, Oslo: International Chamber of Commerce.

World Bank (various years) *World Development Report*, New York: Oxford University Press.

11

ENVIRONMENTAL SOCIOLOGY AND GLOBAL ENVIRONMENTAL CHANGE

A critical assessment

Frederick Buttel and Peter Taylor

INTRODUCTION

The conditions for advance of the subdiscipline of environmental sociology in the early 1990s could hardly be more propitious. Environmentally related issues are more prominent in policy and public discourse across the world than at any time in the history of the subdiscipline. The recent Rio de Janeiro 'Earth Summit' (the UN Conference on Environment and Development) was clearly the largest and most high publicized international conference in the history of the world. Debates prior to, at, and following the Earth Summit on the global conventions on greenhouse gases (and thus on energy and industrial pollution control policies), biodiversity, and forest policy – which, along with stratospheric ozone depletion, have been the leading issues associated with 'global environmental change' – have given these issues an extraordinary amount of public and scholarly visibility.

Despite the remarkable opportunities afforded by this new era of global environmentalism, environmental sociology is not yet making as strong a contribution to understanding global environmental change as a biospheric and sociopolitical phenomenon as it could. There have, in our view, been two distinct, but related limitations of the response of the environmental sociology community to global environmental change. One limitation has been the surprisingly small amount of attention within environmental sociology to the major issues of global environmental change, particularly those such as global warming and stratospheric ozone depletion which have been most closely associated with the rising attention paid to 'global change'. The other limitation is that when sociologists have attended to global change issues, they have tended to do so by uncritically accepting and appropriating the global 'constructions' of modern environmental problems that have emerged within both the environmental sciences and the environmental movement. This is especially problematic since, within both

228

science and politics, the 'globalization' of the environment has served to steer attention to common human interests in environmental conservation, and away from analysing the difficult politics that result from different social groups and nations having highly variegated – if not conflicting – interests in contributing to and alleviating environmental problems (Taylor and Buttel 1992). In particular, as we note later, the environmental sociology community was largely caught unawares by the increasingly bolder Third World opposition to the global climate, biodiversity, and forest management conventions that were prepared for ratification at the 1992 Earth Summit (Pearce 1991a).

While we do not develop this point in this chapter, the modest response of the environmental sociology community to global environmental change is consistent with several traditions within the sub-discipline and its parent discipline. Sociology has long had difficulty conceptualizing global dynamics such as the international state system, international regimes, and the world market. The classical tradition, which remains the basic thrust of modern sociology, has been to take the national-state and national society as the self-evident units of analysis, and to see 'nationally-ordered' problematics (e.g., national class structures, national political processes and policies, and societal-level cultural shifts) as the most important research questions (Sklair 1991). The tradition of environmental sociology since its founding about 20 years ago, in terms of its conceptualization of the environment and environmental dynamics, has paralleled that of the parent discipline. This tradition has been dualistic and of limited applicability to global (or ostensibly global) environmental issues. On one hand, many environmental sociologists have tended to approach environmental dynamics in terms of the specificities of a single, typically localized, environmental problem such as toxic waste contamination (for example, Levine 1982), while employing the community as the level of analysis. On the other, the environment has been conceptualized as a mostly homogeneous, undifferentiated whole in a national framework (for example, Schnaiberg's (1980) notion of societal–environmental dialectic and his implicit contradiction of nationally ordered economic growth vs the environment). Neither approach, however, is well-suited to understanding the multifaceted reality of global change, which involves very complex ecological relations (between locally functioning and globally functioning ecological systems), and complex social–environmental relations (within and between the local, meso, and global levels).

In our view, a more productive role for environmental sociology in addressing issues of global environmental change will require advance on two fronts. The first is that of conceptualizing the mutual relations of causality among the national-state/society, the

229

international state system, and the global economy. As Sklair (1991) and McMichael (1990) have made notable contributions to this end, we do not address this matter here. Second, given the many intersections of science, social change, and politics in global environmental issues, we argue that environmental sociology will need to elaborate an explicit sociology of (environmentally related) science in order to address these issues. We suggest below some promising avenues along which an explicit sociology of science, rooted in the debates over the past 15 years in the field of social studies of science, can be formulated. Before doing so, however, we begin by revisiting the principal vantage point from which environmental sociologists have conceptualized the increasingly stronger environmentalist sentiments within modern societies, and assess the appplicability of this vantage point to understanding the growing concern with global environmental change.

GREENING AND NEW SOCIAL MOVEMENTS

Social change, greening and new social movements

One of the ways that environmental sociologists have made notable contributions to our knowledge on environmentally related phenomena has been through their acumen in recognizing and anticipating social trends and in conceptualizing the emergence of new social forces. One of the most important social trends of our time is 'greening', which sociologists have studied from a variety of angles from the time that it was only an incipient social force. There have been a number of notable attempts to explain the rise of greening on distinctively – though not necessarily exclusively – sociological grounds. One of the major theoretical premises of modern sociology, and the most comprehensive view of the rise of greening to date, concerns the phenomenon of 'new social movements' (NSMs), particularly useful overviews of which are Olofsson (1988) and Scott (1990).[2] The environmental movement and its close cousin, green parties, are typically seen to be the prototypes of new social movements, the rise of which is accounted for by some major structural changes and inertial forms in modern advanced-industrial social structures.

The first and most important force underlying the rise of NSMs has been the progressive demise of the numerical and political position of the traditional industrial working classes in the overall class structures of the advanced countries – from about 45 to 50 per cent of the workforce during the ten years after the Second World War (Singelmann 1978) to less than 20 per cent today (Prezworski 1985, Prezworski and Sprague 1986). The traditional labour, working-class, or social democratic parties thus have increasingly exhibited a structural crisis: they must attempt to accommodate their historic working-class constituency and related ones such as subordinate racial or ethnic minorities (for example, African-Americans in the US). At the same time, the relative size of working-class and related subordinate groups in the electorate has declined to the point that it is too small a base for a winning coalition. Thus, social democratic parties must simultaneously appeal to other classes and groups, to such an extent that non-working-class voters must typically be twofold or more those of its working-class voters in order to win elections. In the main, this means that the social democratic parties must make limited appeals to the middle classes, particularly middle-class liberals among whom (NSM-type) concerns such as feminism, environment, peace/disarmament, or anti-nuclear are particularly salient. Working-class voters, however, have generally not proved to be very vital constituencies of 'new social movements' agendas, while members of the 'new class' are often critical of the centre-left or social democratic parties for so timidly embracing their concerns.

There are two major consequences of the crisis of social democratic parties: (a) As increasingly widespread state fiscal crises have undermined, or threatened to undermine, the traditional social democratic agenda of social Keynesianism, the welfare state and the social wage, many working-class voters desert to the rightist parties. Not enough working-class voters and white-collar liberals stay with the social democratic parties consistently or dependably enough to give them working majorities. This accounts in substantial measure for the pronounced rightward shift in the political centres of gravity of the OECD countries, typified not only by Reaganism and Thatcherism, but also by the growing incidence of Christian Democratic rule even in some of the Northern European Nordic countries, which as recently as a decade ago were thought to be permanently under Social Democratic Party rule; (b) The failure of white collar liberals to achieve their agendas through party politics has led middle- and upper-middle-class activitists to shift their focus to social movements – so-called new social movements – as an alternative to conventional party and parliamentary politics. This means a growing orientation within the 'new class' (and more broadly in the class structure) to membership in and financial support of a growing array of local to national (and, increasingly, international or internationally oriented) environmental, feminist and peace/disarmament organizations that serve as freestanding, usually formally nonpartisan, pressure and lobbying groups, rather than as party constituents. Even where these organizations choose to enter party politics, they do so as self-proclaimed nonconventional or no-business-as-usual parties such as the Grünen of Germany.[3]

To the degree the NSM account is correct, it suggests a structural connection between the neoliberal or conservative drift of the politics of the major industrial nations on one hand, and the rising tide of the increasing bolder expression of green sentiments on the other. Increasingly, it would seem, NSMs have begun to supplant social democratic parties and trade unions as the bulwark of opposition to conservative parties and politics. (See Scott 1990 for a useful discussion of this issue.)

The limits of NSM reasoning

As comprehensive and insightful as the NSM account of the rise of greening is, the NSM perspective none the less has a number of limitations in serving as an orienting posture on the politics of global environmental change. One potential problem with the standard NSM account is that it will tend to see global environmental concerns and mobilization as being (a) merely logical and unproblematic extensions of 'environmental enlightenment', in which bearers of pro-environmental values shift their attention to the international environmental issues that environmental scientists increasingly agree are the more serious ones, and (b) a logical extension of long-standing concerns, such as disarmament and peace, among groups such as the German Grünen.

As we suggest below, however, these views of the globalization of environmental politics and discourse have some important limitations. One is that the 'global construction' of environmental issues is as much a matter of the social construction and politics of knowledge production as it is a straightforward reflection of biophysical reality. The second is that the globalization of environmental policy involves shifts of institutional forums and processes – from national and subnational politics to particular geopolitical arenas such as the international development finance and assistance establishment – that very substantially affect the framing of environmental issues and the consequences of policy decisions.

A second shortcoming of the NSM- and social-values-oriented formulation is that it tends to give short shrift to – and in some cases has misunderstood – the role of scientific knowledge claims and their relationships to NSM movement structure, ideology, and strategy (as Frankel 1987, has noted in a particularly insightful way). NSM theory, of course, recognizes that many ecological issues (for example, industrial toxic wastes, pesticide pollution, land degradation) are derivative of science and technological change. Much of NSM theory, particularly its 'culturalist' variant (see Scott 1990: Ch. 6), has accordingly stressed the anti-science, anti-technology, or anti-technocratic aspect of green movements.

232

However, in our view, science has played a quite different, and arguably more influential and complex, role in the 'greening' trend. One of the major factors that has contributed to the rise of greening has, in fact, been the accumulation of ecological and environmental data and knowledge claims over the past three decades or so, and especially the explosion of global-change-related data and knowledge claims since roughly the mid-1980s (Norgaard 1991a).[4] Further, in contrast to stress in NSM theory on the 'anti-science' and anti-technology undertow of NSM adherents, the rising persuasiveness of environmental and global-change data has contributed to shifting the essential thrust of modern environmentalism towards an increasingly thoroughly 'scientized' *Weltanschauung* and mode of social movement strategy. To be sure, many of the most radical Greens and Earth First!ers distrust all or most 'establishment' science (including academic ecology or environmental science), and prefer to base their claims and agenda on ethical principles rather than mainly on scientific data. None the less, modern environmentalism, where the rubber meets the road, is increasingly an arena characterized by the deployment of scientific and technical knowledge, often in combat with rival data and knowledge claims that are set forth by their industrial, governmental, or quasi-governmental adversaries in an attempt to 'deconstruct' and delegitimate environmental claims (Buttel 1992, Jasanoff 1992, Taylor and Buttel 1992, Yearley 1991). In sum, a more complete account will need to directly consider the sociology of environmental science, which has yet to be grafted on to either the standard 'realist' NSM account of Claus Offe (1987), and especially on to the 'culturalist' account of Touraine (1981).

Another limitation of the standard NSM account of environmentalism is that it has considerable difficulty in explaining Third World environmentalism. Environmental movement organizations have emerged in a number of developing countries over the past half dozen years or so (Adams 1990). But in most Third World contexts, the rise of environmental movements and organizations can hardly be accounted for by either the demise of the mass industrial working class (which they have never had) or a reaction against the mass consumerism of mature industrial society (which they have yet to experience in a thoroughgoing way). In other words, the new social movements account has difficulty understanding greening in Third World nation-states that, in general, are not undergoing the structural or cultural-ideological shifts attributed to the advanced industrial countries. As we suggest below, one cannot account for Third World environmentalism without understanding the social construction and political economy of environmental knowledge – all the way from the laboratory to geopolitical forums such as that of the Earth Summit.

ENVIRONMENTAL SOCIOLOGY'S IMPLICIT SOCIOLOGY OF SCIENCE AND ECOLOGICAL KNOWLEDGE: ITS NATURE AND LIMITS

As suggested earlier, the very nature of environmental sociology is that it involves a sociology of science, at minimum implicitly. In this section we make some brief observations about environmental sociology's approach to science, since we believe these postures and presuppositions vitally affect the work that can be done by sociologists on the phenomena of global environmental change. We make several propositions in this regard:

1 Environmental sociologists are mostly strongly pro-environmental, but in general have little formal training in the environmental sciences (which henceforth we refer to in the broad sense of pertaining to the disciplines of ecology, atmospheric/'planetary' science, conservation biology, ecotoxicology, physical geography, and so on).[5] Sociologists interested in any particular environmentally related issue will thus tend to work from popularized accounts of science written by prominent or publicly visible environmental scientists (for example, N. Myers on tropical deforestation; L.R. Brown on land degradation and declining net primary productivity of ecosystems; C. Sagan and S. Schneider on global climate change; and E.O. Wilson and P.R. Ehrlich on biodiversity destruction).

2 Much of environmental sociology could logically be seen as an implicit sociology of science and technology, since it must inevitably deal with scientific knowledge claims about the nature of environmental problems and recommended solutions. Environmental sociologists are by no means innocent of the 'science and technology connection'. Two of the major groups of American environmental sociologists, those in the relevant sections/divisions of ASA and SSSP, have over the past decade gone so far as to relabel their sections 'environment and technology'. But environmental sociologists, even those whose work has a major focus on science, technology, and technological change, tend to have relatively little background on modern debates in the sociology and social studies of scientific knowledge and technology (some relevant summaries of which are Woolgar 1988, Yearley 1988).

3 In any case, however, environmental sociologists cannot yet obtain all the necessary guidance from the current sociology of science and scientific knowledge. This is the case because, for a variety of reasons internal to this field to be discussed briefly here and at greater length below (see pp. 238–40), there have been two major tendencies in the sociology of science over the past two

decades or so, neither of which has given much attention to environmental science as a subject or provided an adequate framework for addressing such a subject.

Traditionally, most sociologists of science trained in the 1960s or earlier have tended to focus their theoretical and empirical work on the Mertonian sociology of scientists and scientific careers (as in Robert Merton's 1973 classic *The Sociology of Science*, and the related work of Zuckerman 1977, Cole and Cole 1973, or Hagstrom 1975). In this tradition there is little focus on the content of scientific knowledge production, since science is conceptualized as a distinctive institution with a particular normative structure appropriate to uncovering the laws of nature. Accordingly, it is assumed that the content of knowledge production in science will mirror the biophysical parameters of the natural world that scientists must inevitably uncover. Research in this tradition has thus tended to stress the structuring of scientific careers within these normative patterns, and within scientific institutions. Most importantly, science tends to be taken to be analytically demarcatable from society, and to be self-evidently scientific – representing a distinctive combination of formal rationality and value rationality in the Weberian sense.

The other major branch of the sociology of science has been built on a critique of Mertonianism, on the grounds that Mertonianism has served to deny that the character of the knowledge produced by scientists is itself worthy of sociological study. This non- or anti-Mertonian approach was given its original impetus by Thomas Kuhn (1962) in his *The Structure of Scientific Revolutions*, but more recently it has been deepened through recourse to 'interpretive' frameworks such as hermeneutics, symbolic interactionism, ethnomethodology, and cultural sociology. Thus, much of the thrust of the field over the past 15 years has been to invoke various forms of relativism from the sociology of knowledge. In the main, this has meant that the most 'paradigmatically scientific' disciplines, in which one would expect that the Mertonian categories of scientific norms and practices – such as universalism and disinterestedness – would be most strongly manifest, are considered more interesting or stronger test cases for relativism against the more standard Mertonian treatment.[6] If, for example, one can see evidence that scientific knowledge is affected by material interests, social power, rhetorical strategies, and so on in the basic sciences most removed or insulated from the society at large, then a compelling case can be made that these forces operate in all of science. This has meant, in turn, that empirical work in the sociology of science has been focused mainly on the physical sciences, and on the very basic biological sciences such as molecular biology that are either closely connected to the physical sciences or whose subject matter is perceived to be equally insulated from societal forces.

4 The relativist/constructivist turn of the modern sociology of science has been a step forward from the standard Mertonian account, and it has even been reflected in a minor, though yet limited, way in environmental sociology (for example, Dietz 1987, Dietz *et al*. 1992). It is our observation, however, that there has been an overall pattern of selective relativization of scientific knowledge claims in environmental sociology. Environmental sociologists have been at home in relativizing, or demystifying, the knowledge claims that come from anti-environmental quarters such as industrial corporations and industrially funded scientists (for example, as being based on interests, ideology, and the like). They have tended, however, to eschew this type of approach to the environmental sciences in general (Bird 1987), and to the science(s) of global change in particular (Buttel *et al*. 1990).[7] There is, in environmental sociology no less than other branches of social science, a tendency to relativize science one does not like, and to assume that science one likes is self-evidently scientific and valid.

5 As we note at greater length below, social scientists, including but not limited to socioloists, have tended to take at face value, and be largely uncritical of, 'global' notions that have been developed within the environmental science and environmental activist communities (but see Turner *et al*. (1990) for a significant exception). Beginning with the Meadows *et al*. (1972) limits to growth study, and continuing through more recent efforts relating to atmospheric pollution, stratospheric ozone depletion, tropical deforestation, and loss of biodiversity, there has been a tendency within the scientifically oriented sector of the environmental movement (for example, Myers and Myers 1982) and the movement-oriented sector of environmental science (for example, Ehrlich and Wilson 1991) to frame these issues in a supranational framework. Some such issues (for example, greenhouse gas emissions) are actually or potentially intrinsically global – that is, they involve global antecedents, affect globally functioning systems, and may have global social and ecological impacts. However, the 'global' status of other environmental dynamics often insinuated to be components of 'global environmental change' – such as industrial toxic pollution, desertification, and soil erosion – is ambiguous, since many of these are largely localized in their antecedents, social consequences, and

environmental implications (Stern *et al*. 1992, Turner *et al*. 1990). Even more important than this occasionally arbitrary pattern of global labelling are two additional facets of this pattern of social construction and political-economic framing of ecological knowledge: (a) the factors that lead to global constructions of ecological knowledge to be privileged over 'sub-global' frameworks, and (b) the sociopolitical concomitants of complex global-level computer modelling of global environmental phenomena. Each is taken up briefly below (see pp. 242–3, 244–5).

6 There has been a tendency in sociological – and, in general, in social science – circles for there to be premature closure on the 'fact' or 'facts' of global change (especially on the global threats posed by atmospheric warming, tropical rain-forest destruction, and loss of biodiversity). While, for example, there has been a substantial amount of debate and conflicting evidence in the climate and environmental sciences about global environmental change (compare Reifschneider 1989 and Bryson 1990 with Kellogg 1991 and Schneider 1991 on global warming; and Ehrlich and Wilson 1991 and Mann 1991 on biodiversity) and 'scientific uncertainty' is ritualistically acknowledged, sociologists have seldom inquired into the processes by which claims of the likelihood of global catastrophe have been selected for over others (Mol and Spaargaren 1992, Taylor and Buttel 1992). Global warming, in particular, has generally been taken more or less at face value as established scientific fact.[8]

7 Due to premature closure on the stylized facts of global change, there has accordingly been a premature stress on global change problem amelioration, usually as defined *vis-à-vis* the agendas set forth by authoritative spokespeople in the environmentally related sciences and environmental movement organizations. Good examples are Stern *et al*. (1992) and recent special issues of *Evaluation Review* (vol 15, February 1991) and *Policy Studies Journal* (vol. 19, Spring 1991) devoted to global climate change and policy. Research of this type tends to work from the parameter estimates (or some range of estimates), usually as presented by scientifically oriented environmental activists or movement-oriented environmental scientists in a position to speak authoritatively on these issues. In the case of global warming, most of the social science literature begins with the most widely circulated data and projections on global warming (that is, estimates of 4°C or so global warming by the middle of the twenty-first century).[9] Even the most original, critical, and provocative work on global change in the social sciences – that on the distributional and political implications of current and prospective strategies for ameliorating global environmental problems

(see, for example, Kasperson and Dow 1991 in the *Evaluation Review* collection cited earlier) – invariably takes such stylized parameter estimates as its point of departure. Work of this sort is valuable but, as we suggest below, it does not exhaust the range of contributions that environmental sociology can make to global change issues.

TOWARDS A REORIENTATION OF ENVIRONMENTAL SOCIOLOGY SCHOLARSHIP ON GLOBAL CHANGE: SOME ILLUSTRATIVE APPROACHES

Let us now suggest some approaches to diversifying the role of sociology in the study of global change. We begin with some relatively general orientations and then take up some of the specificities of alternative ways to supplement the current approach to global change issues.

Approach 1: Reconsidering environmental sociology's implicit sociology of science and technology

It is useful to begin with an overall view of the (metatheoretical) property space of the sociology of science and technology in order to grapple with how environmental sociologists and others have tended to approach science and technology phenomena. The basis of the typology is two dimensions of science and technology, both of which reflect dualities that are *a priori* reasonable views of the nature of science and technology and the role of science and technology in society.

The first dualism is that which we call 'deference' towards vs 'demystification' of science and technology. A deferential orientation towards science is one in which science is viewed as either intrinsically good, on account of its distinctive decision rules ('rational', 'scientific', 'universalistic', 'disinterested', etc.) that demarcate it from 'nonscience', or as being, in principle, an *a priori* socially desirable activity if organized appropriately or rationally. Demystification of science, by contrast, involves relativizing scientific knowledge claims or scientific accomplishments as being relatively 'ordinary' social constructions, or by being derivative of interests, political-economic relations, class structure, socially defined constraints on discourse, styles of persuasion, and so on. The second dualism is that of science and technology as a social 'practice' or ideational sphere on the one hand, and science and technology as a material-productive force on the other. The typology thus has four cells, with its major categories or exemplars being Mertonian

238

functionalism (deference/practices), relativism/constructivism (demystification/ideational), political economy of science (demystification/material-productive), and a mixed category, consisting of a diversity of approaches such as induced innovation, technology assessment, sociology of risk/risk assessment, and so on (deference/material-productive).

There are two related conclusions about these essential dualisms of science and technology in society. The first is that while each of the major prevailing perspectives in the field – Mertonian-style functionalism, relativism/constructivism, induced innovation, political economy of science and technology – has its role to play, each is incomplete, since each is based in only one quadrant of the typology. The second is that a mature sociology of science and technology, and accordingly a mature environmental sociology, must transcend each of these dualisms so as to straddle each of the quadrants of the typology and incorporate their insights.

It is also important to note for present purposes that for many decades the bulk of the sociology of science has been tilted towards the practices/ideational pole and, in recent years, towards the relativism/constructivism quadrant. By contrast, the bulk of environmental sociology, in so far as it actively considers science and technology, has been tilted towards the material-productive pole, and mostly towards the risk/technology assessment-type mixed approach of the lower left hand quadrant. Further, the approach in most of environmental sociology towards science and technology is usually pursued quite autonomously from the existing literatures in the sociology and social studies of science. It is, as noted in the foregoing, a mostly implicit sociology of science.

There are, of course, some major exceptions to these tendencies. One is the work of Dorothy Nelkin, a scholar whose sociology of science work is cited quite frequently by environmental sociologists, who has eschewed relativism/constructivism and who has focused her work on the premise that science and technology are, first and foremost, components of the material-productive spheres (see, for example, Nelkin 1984). Another exception is the environmental sociologist, Thomas Dietz, some of whose work on environmental risk assessment lies towards the relativism quadrant of the typology (for example, Dietz and Rycroft 1987, Dietz 1987). It is also worth stressing here that Dietz, while a member of a sociology department, has his Ph.D. in environmental studies. He is in a better position than most environmental sociologists to understand the science, and to scrutinize the processes by which environmental knowledge is constructed when this aids sociological explanation. None the less, our typology suggests in schematic terms why, these exceptions

notwithstanding, there is not much cross-fertilization between environmental sociology and the sociology of science, even though their
subject matters obviously overlap to a considerable degree.

While this is not the time and the place to give a full exposition
of what this mature sociology of science and technology might look
like, a few remarks are in order. To the extent that the exponents
of the four different metaperspectives see their work as being conceptually or methodologically in competition then the synthesis must
be more than a simple combination of them. The synthesis we have
been exploring also warrants the name constructionism, but let us
distinguish it from the social constructionism of the upper right
quadrant. To claim that scientific knowledge is constructed is, very
broadly, to say that it is not given by nature. Instead, what counts
as knowledge is contingent on the scientists establishing (or disputing)
it, and through them, on their social context. In exposing the ordinary
quality of the practices of scientists and demystifying the special status
of science, social constructionism emphasizes the malleability of scientific knowledge, practices, and institutions. The synthesis we have in
mind moderates this tendency to relativism, but not by reasserting the
traditional view that the strength of science rests on the corespondence
of scientists' models and theories to natural reality. Instead, we interpret
constructionism to mean that science and politics are co-constructed:
that is, scientific accounts are difficult to modify to the extent that they
facilitate and are, in turn, facilitated by favoured social policies, actions
and interventions (Taylor 1992, Taylor and Buttel 1992). This proposition about 'action-oriented' constructionism derives from the following
observations: a scientist's accounts can be accepted or disputed by many
different agents. Scientists seek to ensure their work is promoted rather
than discounted by these agents by mobilizing diverse resources –
categories, equipment, data, experimental protocols, citations,
colleagues, the reputation of research institutions, rhetorical devices,
funding, media publicity. In doing so, technical and social considerations tend to reinforce each other, that is, theories and actions render
each other more difficult to modify *in practice*. (See Taylor 1992, 1993,
for elaboration on this necessarily condensed exposition.)

The two dualisms then become special emphases within this action-
oriented constructionism, varying according to who is trying to modify
some science/technology, who resists them, and the divergent
resources that are exposed in the process. In contrast, a mature
sociology of science and technology – and, accordingly, a sound
environmental sociology – will have a more diversified approach. It
will give more stress to the ideational or 'practices' components of
these phenomena, but without sacrificing its entirely justifiable stress
on science and technology as material-productive forces.

Approach 2: Exploring the relationships between environmental science/concepts and ideologies/movements: the construction and deconstruction of global environmental knowledge

It is arguably the case that environmental sociology has been limited by confining its attention to phenomena outside of the laboratory, while the modern sociology of science, particularly its dominant relativist/constructivist wing, has been limited by giving predominant stress to knowledge at the laboratory level (Latour 1987, Cozzens and Gieryn 1990). One of the most promising foci for both subdisciplines for transcending their limitations is that of the environmental sciences and environmentalism (Taylor 1991).

Global change is a good example of a construction that serves simultaneously as a scientific concept (and knowledge claim) and as a movement ideology (of environmentally related movements). This is by no means a novel circumstance; modern environmentalism in its early years was undergirded by the notion of the 'population bomb' (a derivative of population biology) *à la* P.R. Ehrlich (1968), and later by the 'limits to growth' (derived from the application of system-dynamics to ecological systems) *à la* Meadows *et al.* (1972). The ascension of global change as the predominant ideology of the environmental movement (particularly within its dominant, internationally oriented wing) thus reflects a more long-standing trend of the 'scientization' of environmentalism and NSMs.

As suggested earlier, the environmental sciences and environmental movements, both broadly construed, exist in a state of mutual dependency and contradiction. At the most general level, the environmental movement depends on persuasive environmental science knowledge claims, and the environmental sciences stand to benefit substantially from a politically persuasive environmental movement.

The environmentalist/environmental science relationship is revealed further in the interrelated roles that environmental scientists and activists increasingly play. One such role, alluded to earlier, is that in which environmental scientists engage in activism. A related one is that in which environmentalists take on the role of 'quasi-scientists'. Increasingly, these roles are being collapsed. For example, contemporary environmental organizations are increasingly staffed with Ph.D. holders who have scientific titles, and often résumés that resemble those of academics. Persons such as S. Schneider and N. Myers are good examples of the emerging ideal type of scientist–activist.

Despite the tendency to complementarity between the environmental movement and environmental science, this cannot explain why particular kinds of environmental knowledge claims – in particular,

241

ones positing a global-level dynamic and constructed at a global level of analysis – will tend to be privileged over others. We believe there are two particularly important social forces, one within environmentalism and the other within environmental science, that must be considered in accounting for these dynamics. The first, which we have treated elsewhere (Taylor and Buttel 1992) and so only mention here, is the tendency within environmental science for global formulations to be privileged or selected over others. The second has to do with the constraints on environmental mobilization and political action.

Global change/warming has proved to be very attractive to the international environmental movement because of two imperatives, both intrinsic to environmental issues, with which the movement must deal. First, in so far as environmental goals tend to be public goods, 'saving' the environment is in everyone's interest, and hence no one's in particular, leading to a potentially very difficult collective mobilization problem. Second, the environmental agenda, as a disparate congeries of specific issues (for example, reducing pollution, conserving biodiversity and natural habitats, conserving natural resources, promoting recycling, increasing the efficiency of energy utilization), tends to involve too many forums, too many battles, too many conflicting interests, and too many opponents to be realistically achievable. Global formulations readily lend themselves to dealing with the problems of environmental mobilization and multiple policy forums. Global (scientific) formulations permit 'packaging' of multiple environmental problems and concerns within a common, overarching rubric, at the same time that they convey the legitimacy and persuasiveness afforded by their being rooted in science. Formulations that provide scientific justification for world-wide 'alarm' or 'dread' are particularly attractive in both authenticating this 'packaging' approach, as well as in creating the political rationale for responding urgently (Mol and Spaargaren 1992). The creation of a supranational climate (no pun intended) of urgency in responding to humanity- and biosphere-threatening problems enables the movement to make authoritative moral and ethical claims that it is imperative for all groups to co-operate in overriding the politics-as-usual associated with the multiple local, regional, and national forums in which pro-environmental policies would otherwise need to be pursued. The international political pressures that led to the two 1970s UN environmental conferences (on environment and population) and to the 1980s World Commission on Environment and Development (see WCED 1987), and to the conventions and treaties prepared for ratification at the 1992 Earth Summit, are cases in point.

Global change/warming, much like the 'population bomb' and the 'limits to growth' in previous decades, became plausible as a consolidating framework, and thus as an overarching movement ideology, for several reasons. In addition to the legitimacy afforded by its scientific imprimatur, global change was particularly appropriate for aggregating the bulk of the traditional environmental agenda (for example, industrial pollution control, energy conservation, preservation of tropical ecosystems, population control) under a single umbrella and rationale. Global change/warming lent itself to popularization[10] on account of multiple projected 'dread factors': massive coastal inundation due to rising sea-levels, increases in cancer due to ozone layer depletion, destructive impacts of climate alterations on agricultural productivity (especially in the American and other temperate breadbaskets), a growing incidence of drought and climatic extremes, the spectre of wholesale loss of biodiversity, and so on. These dread factors, or what Mol and Spaargaren (1992) have called 'eco-alarmism', were integral in constructing a portrait of global change in which it was stressed that communities, regions, and nations are impotent to deal with these problems on their own – hence the need to override 'politics-as-usual' and urgently to erect a new global regulatory order with the moral imperative to address these profound threats to human survival and biospheric integrity.

The recent tendency towards fusion or convergence of the roles of environmental activist and scientist notwithstanding, the mutual dependency of movements and science may also be conflictual or contradictory. The environmental movement, for example, has long tended to regard environmental scientists as being too timid in bringing their findings to the public and in taking political stands on environmental issues. Scientists likewise may become uncomfortable when faced with the transparency of the ideological moorings or implications of their research. One such example, in the case of global warming, was the tendency that emerged among some climate/planetary scientists, beginning in late 1989 and early 1990, to express ambivalence about the political uses of their data, and to distance themselves from movement ideology and the more radical fringes of green forces (Buttel *et al.* 1990).

As much as global change/warming was successful as an environmental ideology and mobilization strategy, it must be recognized as well that movements that base their claims and agenda on scientific knowledge claims are also vulnerable to their ideas being 'deconstructed' by and through science. 'Scientific uncertainty' can be an enormously powerful tool, and is one that is often wielded against environmentalists with particular effectiveness (Jasanoff 1992). In so far as the perception of scientific consensus about the likelihood of the 'greenhouse effect' was one with which not all climate/planetary scientists were entirely comfortable, it was almost inevitable that there would emerge scientific studies and opinion that would cast doubt on the portrait painted by environmentalists. Accordingly, the work of scientists who resisted or were agnostic about the conclusions or policies advocated by proponents of global warming – on either ideological (Marshall Institute 1989) or more conventionally scientific grounds (Bryson 1990) – would ultimately play a significant role in global warming politics. While active opposition to policies to ameliorate global warming is currently confined largely to the Bush Administration and to a number of governments and activists (as well as tropical timber entrepreneurs) in the Third World, the scientific data and opinions that contradict the environmentalist rendering of global climate data are playing a very significant and growing role in justifying opposition to 'greenhouse policies' (useful examples of which in the mainstream media are Nordhaus 1990a, 1990b; *The Economist* 1991).

Approach 3: Global change mobilization in a context of free-market resurgence

We noted earlier some of the limitations of NSM-type value-oriented theories as an orienting perspective on global environmental change. Here it is useful to extend these observations by noting that NSM-type theories, by focusing on the politics of the environment from the vantage point of broad social values rather than institutional structures, may exaggerate the degree to which NSMs are a potent oppositional force in modern politics or in geopolitics. In particular, NSM perspectives alone have difficulty accounting for two interrelated political realities of the current era of global environmental change. There is a seeming contradiction between the ostensible radicalness and oppositional character of NSM ideologies and two current phenomena: (a) global environmental change has yet to prompt concerted attempts at corporate veto; and (b) environmental movement organizations have not only accommodated themselves to the free-market resurgence of the 1980s and 1990s, but some of them have aligned themselves with the dominant institutions of global society (particularly the international development finance and monetary establishment), often against the (immediate) interests of groups and nations of the South.

It is useful to begin by noting that the limits to growth, which has many similarities to global change,[11] was in substantial measure delegitimated almost from the start through corporate veto. Implementation of a limits-to-growth world view would have severely constrained capital accumulation, would have virtually required a nationally planned economy and sharply increased state intervention, and would have threatened those whose interests were tied to growth.

Global change, which has been widely popularized since the summer of 1988, has attracted little scrutiny of this sort. To our knowledge, the only significant advanced-country corporate opposition to ongoing attempts to forge a climate change convention (prepared for ratification at the Earth Summit, ultimately in a very diluted form) has been that of the Climate Council, a US lobbying group representing electrical utilities that depend heavily on fossil fuels, particularly coal (Pearce 1991b).

To some extent the puzzle of the lack of intense corporate opposition to, or attempts at corporate veto of, global warming policy can be explained by the business opportunities that currently popular responses to global environmental change will afford. These policies will, in particular, involve the likelihood that carbon taxes will be the centre-piece of a global climate convention. Carbon taxes could lead to the revitalization of the civilian nuclear power industry as a means of providing growing levels of energy with lower levels of CO_2 emissions (*The Economist* 1989a). Another area of considerable profit potential is that of R&D into new industrial chemicals and plant-based biotechnology processes that can substitute for petroleum-based production of CFCs and other chemicals (*The Economist* 1989b).[12]

Over and above the new R&D and commercial opportunities, the lack of corporate opposition to the global climate convention probably stems from the green world view and environmentalist strategy themselves. At one level, as we have observed elsewhere (Buttel *et al*. 1990, Taylor and Buttel 1992), global change was promoted in a selective way, so as to generate support among prospective environmental supporters and to minimize opposition among the political and corporate officialdoms in the advanced industrial countries.[13] At another, modern environmentalism has accommodated itself surprisingly readily to the global free-market resurgence. While international environmental groups yet reserve the right to criticize the World Bank and related institutions about the environmental destruction that results from particular projects or types of projects (especially dam, road construction, and mining projects; see Lewis 1991, Hunt and Sattaur 1991), environmental groups have generally worked with the Bank in a surprisingly harmonious manner in implementing conservation/preservation policies and programmes in the Third World (Parker 1991). There is a key coincidence of interest in the environmental group/World Bank/IMF relationship: the Bank and IMF gain legitimacy in the eyes of the citizens and political officialdoms of the advanced countries by helping to implement environmental and conservation policies, while the implied threat of Bank or IMF termination of bridging, adjustment, and project loans is useful in securing developing-country compliance with environmental initiatives.

Given the relative harmony of this relationship, the environmental community has been disinclined to take on the world debt crisis, the net South-to-North capital drain, and the international monetary order (which is substantially regulated by the World Bank and IMF; see Wood 1986) as being fundamental contributors to global environmental degradation. In a breaking of ranks that is the exception that proves the rule, Postel and Flavin of the Worldwatch Institute have recently (1991) stated that the South-to-North capital drain (now approaching US$50 billion annually) and the environmentally destructive imperatives of Bank- and IMF-supervised debt repayment must be addressed before permanent solutions to global environmental problems can be implemented.[14]

Approach 4: Turning up the heat: global change, development discontent, the 'debt connection', and the road to and from Rio de Janeiro

It is increasingly widely recognized that the popularization of the global warming notion was accompanied by, if not substantially based on, giving disproportionate stress to Third World sources of greenhouse gases, particularly tropical rain-forest destruction. Tropical rain-forest destruction, however, probably accounts for less than 15 per cent of global greenhouse gases (Norgaard 1991b), and is a relatively minor source compared to industrial, transport, and other greenhouse gas emissions from the developed countries.[15] It is, of course, likely that if the ambitious energy and overall development plans of developing countries (particularly China) are implemented, there will be a very considerable expansion of their greenhouse gas emissions over the next few decades. None the less, the arguably disproportionate stress given to the rain-forest component of global climate change has been among the major catalysts of developing-country opposition to a global climate treaty that was prepared for ratification at the Earth Summit (Pearce 1991a).

Another source of North–South friction over the global climate and other conventions prepared for ratification at the Earth Summit is more long-standing. International development policy has long been conflictual, involving struggles between official development agencies and external groups critical of the performance and consequences of development projects and policies. One of the most important, yet largely invisible, concomitants of the conservative drift of Western politics has been the implementation of 'structural adjustment' doctrine within international development finance and assistance institutions, most notably the World Bank and IMF. The main impetus for the structural adjustment policies that have been imposed on developing countries has been the global debt crisis, and the resulting international monetary instability, which while nearly 10 years old is no closer to resolution than when Mexico first defaulted on its loan payments in the early 1980s (Canak 1989). The political economy of debt has become the principal parameter affecting Third World development prospects. Most important for present purposes, the debt crisis – and the structural adjustment policies imposed on the Third World in order to extract as much interest and principal as possible and to sustain the belief that the Third World debt will ultimately be repaid – has been accompanied by the undermining of the traditional means by which opposition to official development policies had been articulated by many developing-country governments and most development activists. Social-justice-based opposition to development policies – on the grounds that these policies are unwarranted because they aggravate inequality or fail to improve the lot of the poor – has increasingly lost its standing and the influence it once wielded in institutional forums such as the World Bank/IMF, the US Agency for International Development, and the UN system. Increasingly, in this era of debt crisis and structural adjustment, environmental criticism of development policies and projects now serves as the predominant discourse for expressing opposition to official development policy.

The process of substituting environmental for social justice discourse, however, is contradictory. It has largely been through the 'debt regime' that environmental agendas have been grafted on to Third World development planning. Only heavily indebted countries, for example, have debt that is sufficiently discounted on the secondary debt market to be attractive to environmental groups for purchase in debt-for-nature swaps. Likewise, heavily indebted countries are most subject to joint environmental and development agency pressures to protect the environment. But as much as external debt has facilitated the implementation of environmental conservation policies, debt also serves to exacerbate environmental degradation. Third World countries that are most 'debt-stressed', and thus who are most in need of hard-currency export revenues, are most likely to see little alternative but aggressively to 'develop' their tropical rain forests and other sensitive habitats in order to maintain their balance of payments and service their debts.

It is therefore not surprising that there has emerged a growing Third World reaction to 'environmental colonialism'. This reaction is surprisingly broadly based within the developing world. Much of its intellectual rationale has been articulated by left-leaning groups such as the Centre for Science and the Environment in India. These groups have stressed that international environmental organizations have exaggerated the Third World contribution to global warming, and that Western calculations of developing country contributions to greenhouse gas emissions have failed to note a fundamental First World/Third World difference in the nature of these emissions: that between the 'survival emissions' of the 'South' and the 'luxury emissions' of the 'North' (see Agarwal and Narain 1991). But Third World criticism of global environmental policies' environmental colonialism also includes increasingly forceful opposition to proposed global change conventions by Third World politicians and business leaders – not only on grounds of 'national sovereignty', but also through demands that these conventions include binding commitments by the North to subsidize the South's transition to environmental protection through massive expansion of foreign aid (Pearce 1991a). The Earth Summit was accordingly dominated more by the saga of North–South acrimony than by environmental science.

The experiences of the 1970s UN conferences on population and environment as well as that of 1992 Earth Summit suggest that environmental sociology, particularly that which seeks to understand global environmental change, should be reconstructed by giving more attention to international political economy. For example, the two most fundamental institutions of global society today – the inter-state system, based on the principle of state sovereignty; and the world economy, based on GATT and related rules and on international monetary deregulation that sanctions international competition through world trade and through internationalization of finance via floating exchange rates – are largely adverse for environmental protection. As we have seen at the Earth Summit, national-state sovereignty can be and is construed to include the national right to exploit resources at the discretion of their regimes. Also, as seen at Rio de Janeiro, the international competition dynamic may – in ideology, if not reality – compel states to compete effectively with one another by degrading the environment. Another possible lesson from the Earth Summit was the potent reminder of the fact that foreign aid was essentially a product of the Cold War, in which East and West vied for the hearts, minds, and security allegiance of Third World nation-states. Now that the Cold War is over, the impulse to assist the Third World (either through development grants – rather than through the World Bank/IMF loan apparatus and extension of the 'debt trap'; see Canak 1989 – or through major foreign aid programmes to subsidize Third World environmental responsibility) is weak. This fundamental reality – that the bulk of the industrial states now see little geopolitical reason for restoring foreign aid programmes to their 1970s levels, let alone for undertaking major infusions of 'fresh money' for environmental assistance that confers little geopolitical benefit – none the less came as a shock to international environmental activists and Third World officialdoms at Rio de Janeiro. It should be stressed, however, that as crucial as these two principles of global society are in shaping the international political economy of the environment, each was largely laid down during the post-Second World War period and neither is necessarily permanent. Globalization of international finance and of commodity markets is beginning to erode state sovereignty, particularly the ability of states to implement fiscal and monetary policies to achieve national goals without risking national decline in a competitive international economic environment. For many countries, even rich ones with world-market advantages, protectionism and reassertion of the integrity of national economies may in the future be seen as preferable to depending on the vagaries of national economic competition in international markets in money and goods to improve their living standards (Gilpin 1987). The reluctance of world states in coming to agreement on the liberalization initiatives of the long Uruguay round of the GATT negotiations, along with the rising share of world trade that is 'administered', attests to the reversibility of the key post-war institutions of global society. As these dynamics unfold, they will have enormous implications for the environment and for how these problems can and must be dealt with.

CONCLUSION

Neither an environmental sociology which fails to attend to the social construction of environmental knowledge nor a sociology of science that ignores the material-productive realities of environmental knowledge can understand the significance of global environmental change in the world today. We have sought to chart some new directions to this end. Our suggestions have stressed the causes and consequences of the fact that global environmental change serves simultaneously as scientific concept and social ideology, and the utility of identifying the mobilizational and political continuities between global change and previous (global) conceptualizations of environmental issues. We believe that further progress requires more attention to be paid to understanding both the social and political-economic forces that affect the construction of environmentally related scientific knowledge, and the 'scientization' and 'scientific deconstruction' dynamics within environmentalism. We also need to recognize that the internationalization of environmentalism has involved it being shifted towards and grafted on to a set of geopolitical institutions – particularly those of inter-state relations, the world economy and rules of world trade, the development assistance establishment, and implicitly those of the world monetary order – that both decisively shape environmentalism and define its limits in the late twentieth century.

NOTES

1 An earlier version of this chapter was presented at the seminar series of the Program on Social Analysis of Environmental Change, Cornell University, April 1991, and published in *Society and Natural Resources* 5: 211–30, 1992. The senior author's research was supported by funds from a grant from the Environmental Protection Agency to the Ecosystems Research Center, Cornell University.

2 It is by no means the case that all or most environmental sociologists explicitly embrace theories of new social movements. These theories, however, may be seen to be consistent with or to subsume a number of other perspectives (for example, of the rise of postmaterial values, the new ecological paradigm). See Cotgrove (1982) for a noteworthy early attempt to demonstrate how a variety of strands of research in environmental sociology could be subsumed within the (then) emerging European traditions of NSM research, and Buttel (1992) for suggestions as to how NSM theory can be extended to be relevant to global environmental phenomena.

3 It should be stressed, of course, that just as the post-Second World War trajectory of social democratic modernity' and 'Fordism' were manifest differentially and unevenly among the advanced countries (Lash and Urry 1987, Esping-Andersen 1990), the demise of the post-war order has equally diverse manifestations. The portrait as painted above – of the decline of party politics, the shaky coexistence of left parties' working class and 'new class' constituencies, the search among the left to social movement alternatives to parties – pertains most clearly to the Anglo *laissez-faire* regime types (Esping-Andersen 1990). But even in the Nordic nations where the welfare state is largely intact, slow growth, state fiscal crises, and Social Democratic Party disarray have led to Christian Democratic rule and/or moves to neoliberal policies. The rise of NSMs as an alternative to party politics has also been a general concomitant of the disintegration of the post-war order. Also note that there is a disagreement within the NSM literature on whether the predominant thrust of these movements is to serve as a vehicle for expression of 'identity', or whether these movements should be seen mainly as a more clearly instrumentally rational pursuit of the policies deriving from movement ideology (see, for example, Scott 1990; Ch. 6). The former ('culturalist') position is most closely associated with Alain Touraine (1981), and the latter ('realist') position with Claus Offe (1987).

4 This does not mean that modern environmental movements whose ideologies are rooted in science will not put their own gloss on the scientific data used to buttress their claims. For example, while much of the 'sustainable development' agenda, for example, as set forth in the Brundtland Report (WCED 1987), was premised on ecological science, Timberlake (1989) notes that the ideas that underlay the concept of sustainable development were more matters of opinion than scientifically based ones. We note below (see pp. 242–4) how this process has occurred in the popular construction of global climate knowledge.

5 But see Norgaard's (1991) useful discussion of the fragmentation of the environmental sciences and of the implications of the lack of 'a meta-model to link the individual environmental sciences into a coherent whole'.

6 In addition to the preoccupation of the now-dominant wing of the sociology of science with challenging the Mertonian view, many in this wing were as (or more) motivated to challenge Mannheim's sociology of knowledge. Mannheim claimed that his sociology of knowledge was applicable only to the social sciences, since knowledge in the natural sciences would be determined mainly by the laws of nature that they would inevitably uncover. The constructivist wing in the sociology of science, by contrast, emerged, contra-Mannheim, with a conviction that natural science knowledge, no less than that in the social sciences, is socially constructed.

7 The major exceptions to this generalization are several historians of science conversant with the sociology of science (see, for example, Bird 1987). Bramell's (1989) history of twentieth-century ecology, in both the scientific and movement senses, is a particularly notable exception.

8 Note that while there has been criticism of the stylized world view of greenhouse effect proponents over the past year or so, the criticisms by economists (for example, Nordhaus 1990a, 1990b; the work of the Adaptation panel of the NRC/NAS study on global climate change) have been more influential than those of critics within the atmospheric/ planetary science research community. Interestingly, these criticisms by economists and other 'adaptationists' typically take the received parameters of greenhouse warming as their point of departure, while arguing that incremental economic, migration, technological, and other adaptive mechanisms will be sufficient to deal with very slow increases in global mean temperature.

9 As noted earlier, sociologists as a whole have difficulty conceptualizing the 'global' in a distinctly sociological sense. Sklair's (1991) theory of economic, political, and cultural-ideological 'transnational practices', however, is a promising means of treating global dynamics without world-economic or geopolitical reductionism.

10 As important as the modern sociology of science's emphasis on the social construction of scientific knowledge at the laboratory level is (or can be) to understanding environmental data, the rise of global change/ warming also demands attention to the processes of social construction of popular knowledge (see Buttel *et al*. 1990).

11 An earlier publication (Buttel *et al*. 1990) has explored the many similarities of these two frameworks (for example, that both are globally and computer-modelling based, neo-Malthusian, critical of industrial civilization) as well as their differences (for example, that while the LTG saw fossil-fuel scarcity as a major problem, the abundance of fossil fuels is seen as problematic from a global change point of view). See Taylor and Buttel (1992).

12 These substitutes, however, will be highly profitable and can achieve global market penetration only if there is effective global regulation – for example, global phaseouts of the use of (cheaper) CFCs, an effective international convention on climate change. Thus, many of the business opportunities that will be afforded by responding to global climate change are not only consistent with, but will require, international environmental regulation.

13 In particular, the popularization of global warming tended to stress the need to shift to alternative energy sources over the need for strict energy-use reduction, and also gave disproportionate stress to Third World sources of greenhouse gases.

14 We have undoubtedly overgeneralized here. In the early autumn of 1992, as we completed revisions on this manuscript, an international meeting of environmentalists in Washington discussed issues relating to structural adjustment and the environment. The resistance of the Bush Administration of the United States to the treaties prepared for signing at the Rio Earth Summit, together with the influence on the treaty formulation of corporate NGOs such as the Business Council for Sustainable Development, indicates that the contradictions are being brought into sharper relief. See Hecht and Cockburn (1992), Bidwai (1992), and the July/August edition of the journal *Multinational Monitor*, 'Report from Rio', for a more complex account of the conflicts and alliances among environmentalists, business, and First and Third World States.

15 Thrupp (1991) stresses that the pre-climate-change preoccupations of environmental groups (preserving primary rain forests and other 'sensitive' zones, protection of wildlife species) were instrumental in the formulation of environmentalist doctrine on global change. See also Buttel *et al.* (1991) for a discussion of the role of rising consciousness about global climate change in galvanizing the sustainable development movement in the late 1980s and of the tendency to 'rainforest fundamentalism' in sustainable development practice.

REFERENCES

Adams, W.M. (1990) *Green Development*, London: Routledge.

Agarwal, A., and Narain, S. (1991) 'Global warming in an unequal world: a case of environmental colonialism', *Earth Island Journal*, Spring: 39–40.

Bidwai, P. (1992) 'North vs. South on pollution', *The Nation* 254 (22 June): 853–4.

Bird, E.A.R. (1987) 'The social construction of nature: theoretical approaches to the history of environmental problems', *Environmental Review* 11: 255–64.

Bramwell, A. (1989) *Ecology in the 20th Century*, New Haven: Yale University Press.

Bryson, R.A. (1990) 'Will there be a global "greenhouse warming"?', *Environmental Conservation* 17: 97–9.

Buttel, F.H. (1992) 'Environmentalization: origins, processes, and implications for rural social change', *Rural Sociology* 57: 1–27.

—— Hawkins, A. and Power, A.G. (1990) 'From limits to growth to global change: contrasts and contradictions in the evolution of environmental science and ideology', *Global Environmental Change* 1: 57–66.

—— Sunderlin, W. and Belsky, J. (1991) 'Balancing biodiversity and human welfare', Report prepared for the United Nations Research Institute for Social Development, Geneva.

Canak, W.L. (ed.) (1989) *Lost Promises*, Boulder, Colo.: Westview Press.

Cole, J.R. and Cole, S. (1973) *Social Stratification in Science*, Chicago: University of Chicago Press.

Cotgrove, S. (1982) *Catastrophe or Cornucopia*, New York: Wiley.

Cozzens, S.E. and Gieryn, T.F. (eds) (1990) *Theories of Science in Society*, Bloomington: Indiana University Press.

Dietz, T. (1987) 'Theory and method in social impact assessment', *Sociological Inquiry* 57: 54–69.

—— and Rycroft, R.W. (1987) *The Risk Professionals*, New York: Russell Sage Foundation.

—— Frey, R.S. and Rosa, E. (1992) 'Risk, technology and society', in R.E. Dunlap and W. Michelson (eds), *Handbook of Environmental Sociology*. Westport, Conn.: Greenwood Press.

The Economist (1989a) 'Green electricity', *The Economist*, 2 December: 17–18.

—— (1989b) 'Friendly gases', *The Economist*, 20 May: 74, 79.

—— (1991) 'Species galore: avoiding extinctions should not be an overriding goal for environmentalists', *The Economist*, 14 September: 17.

Ehrlich, P.R. (1968) *The Population Bomb*, New York: Ballantine.

Ehrlich, P.R. and Wilson, E.O. (1991) 'Biodiversity studies: science and policy', *Science* 253: 758–62.

Esping-Andersen, G. (1990) *Three Worlds of Welfare Capitalism*, Princeton: Princeton University Press.

Frankel, B. (1987) *The Post-Industrial Utopians*, Madison: University of Wisconsin Press.

Gilpin, R. (1987) *The Political Economy of International Relations*, Princeton: Princeton University Press.

Hagstrom, W.O. (1975) *The Scientific Community*, Carbondale: Southern Illinois University Press.

Hecht, S. and Cockburn, A. (1992) 'Rhetoric and reality in Rio', *The Nation* 254 (22 June): 848–53.

Hunt, P. and Sattaur, O. (1991) 'World Bank's conservation record under fire', *New Scientist* 132 (19 October): 14.

Jasanoff, S. (1992) 'Science, politics, and the renegotiation of expertise at EPA', *Osiris* (new series), vol. 8.

Kasperson, R.E. and Dow, K.M. (1991) 'Environmental and geographic equity in global environmental change: a framework for analysis', *Evaluation Review* 15: 149–71.

Kellogg, W.W. (1991) 'Response to skeptics of global warming', *Bulletin of the American Meteorological Society* 72: 499–511.

Kuhn, T. (1962) *The Structure of Scientific Revolutions*, Chicago: University of Chicago Press.

Lash, S. and Urry, J. (1987) *The End of Organized Capitalism*, Madison: University of Wisconsin Press.

Latour, B. (1987) *Science in Action*, Cambridge, Mass.: Harvard University Press.

Levine, A. (1982) *Love Canal*, Lexington, Mass.: Lexington Books.

Lewis, D. (1991) 'G7 gets a roasting on environmental record', *New Scientist* (20 July): 14.

McMichael, P. (1990) 'Incorporating comparison within a world-historical perspective: an alternative comparative method', *American Sociological Review* 55: 385–97.

Mann, C.C. (1991) 'Extinction: are ecologists crying wolf?', *Science* 253: 736–8.

Marshall Institute (1989) *Scientific Perspectives on the Greenhouse Problem*, Washington, DC: Marshall Institute.

Meadows, D.H., Meadows, D.L., Randers, J. and Behrens, W.W., III (1972) *The Limits to Growth*, New York: Universe.

Merton, R.K. (1973) *The Sociology of Science*, Chicago: University of Chicago Press.

Mol, A.P.J. and Spaargaren, G. (1992) 'Environment, modernity, and the risk-society: the apocalyptic horizon of environmental reform', Paper presented at the Symposium on Current Developments in Environmental Sociology, Woudshoten, The Netherlands, June.

Myers, N. and Myers, D. (1982) 'From the "duck pond" to the global commons: increasing awareness of the supranational nature of emerging environmental issues'', *Ambio* 11: 98–201.

Nelkin, D. (ed.) (1984) *Controversy*, Beverly Hills, Calif.: Sage.

Nordhaus, W.D. (1990a) 'Greenhouse economics: count before you leap', *The Economist* 316 (7 July): 21–4.

—— (1990b) 'To curb or not to curb: the economics of the greenhouse effect', Paper presented at the annual meeting of the American Association for the Advancement of Science, New Orleans, February.

Norgaard, R.B. (1991a) 'Environmental science as a social process', *Environmental Monitoring and Assessment* 16: 1–16

—— (1991b) 'Sustainability as intergenerational equity: the challenge to economic thought and practice', Discussion Paper, Asia Regional Series, World Bank, Washington, DC, June.

Offe, C. (1987) 'Changing boundaries of institutional politics: social movements since the 1960s, p. 63–106 in C.S. Maier (ed.), *Changing Boundaries of the Political*, Cambridge: Cambridge University Press.

Oloffson, G. (1988) 'After the working-class movement? An essay on what's "new" and what's "social" in new social movements', *Acta Sociologica* 31: 15–34.

Parker, P. (1991) 'The advance of the green guards', *New Scientist* (3 August): 44.

Pearce, F (1991a) 'North–South rift bars path to summit', *New Scientist* 132 23 November: 20–1.

—— (1991b) 'Industry scorns climate predictions', *New Scientist* 132 (2 November): 10.

Postel, S. and Flavin, C. (1991) 'Reshaping the global economy', pp. 170–88 in L.R. Brown *et al.*, *State of the World*, New York: Norton.

Przeworski, A. (1985) *Capitalism and Social Democracy*, New York: Cambridge University Press.

—— and Sprague, J. (1986) *Paper Stones: A History of Electoral Socialism*, Chicago: University of Chicago Press.

Reifschneider, W.E. (1989) 'A tale of ten fallacies: the skeptical inquirer's view of the carbon dioxide/climate controversy', *Agricultural and Forest Meteorology* 47: 349–71.

Schnaiberg, A. (1980) *The Environment*, New York: Oxford University Press.

Schneider, S. (1991) 'Three reports of the Intergovernment Panel on Climate Change', *Environment* 33: 25–30.

Scott, A. (1990) *Ideology and the New Social Movements*, London: Unwin Hyman.

Singlemann, J. (1978) *From Agriculture to Services*, Beverly Hills, Calif.: Sage.

Sklair, L. (1991) *Sociology of the Global System*, Baltimore, Md.: Johns Hopkins University Press.

Stem, P.C., Young, O.R. and Druckman, D. (eds) (1992) *Global Environmental Change: Understanding the Human Dimensions*, Washington, DC: National Academy Press.

Taylor, P.J. (1991) 'Environmental studies as a key research site for social studies of science', Paper presented at the annual meeting of the Society for Social Studies of Science, Cambridge, Mass., November.

—— (1992) 'Re/constructing socio-ecologies: system dynamics modelling of nomadic pastoralists in sub-Saharan Africa', pp. 115–48 in A. Clarke and J. Fujimura (eds), *The Right Tool for the Job: At Work in the Twentieth Century Life Sciences*, Princeton: Princeton University Press.

—— (1993) 'Building on the metaphor of construction in analytic science studies', Manuscript submitted to Social Studies of Science.

—— and Buttel, F.H. (1992) 'How do we know we have global environmental problems? Science and the globalization of environmental discourse', *Geoforum*, vol. 23.

Thrupp, L. (1991) 'Politics of the sustainable development crusade: from elite protectionism to social justice in Third World resource issues', *Environment, Technology, and Society* 58: 73–91.

Timberlake, L. (1989) 'The role of scientific knowledge in drawing up the Brundtland Report', pp. 117–23 in S. Andersen and W. Ostreng (eds), *International Resource Management*, London: Belhaven Press.

Touraine, A. (1981) *The Voice and the Eye: An Analysis of Social Movements*, Cambridge: Cambridge University Press.

Turner, B.L. *et al.* (1990) 'Two types of global environmental change: definitional and spatial-scale issues in their human dimensions', *Global Environmental Change* 1: 14–22.

Woolgar, S. (1988) *Science: The Very Idea*, London: Tavistock.

Yearley, S. (1988) *Science, Technology and Social Change*, London: Unwin Hyman.

—— (1991) *The Green Case*, London: HarperCollins.

WCED (World Commission on Environment and Development) (1987) *Our Common Future*, New York: Oxford University Press.

Wood, R.E. (1986) *From Marshall Plan to Debt Crisis*, Berkeley: University of California Press.

Zuckerman, H. (1977) *The Scientific Elite*, New York: Free Press.

255

12

SUSTAINING DEVELOPMENTS IN ENVIRONMENTAL SOCIOLOGY

Elizabeth Shove

'A flash in the pan', said a colleague, 'a passing fashion but nothing really to do with sociology.' It is easy to dismiss this assessment of environmental sociology without further thought and there may be good political reasons for taking the importance of our subject for granted. On the other hand, the sceptics may have a point. Could concern about the environment 'blow over' as they anticipate? The range of issues and approaches now gathering under the environmental banner is certainly impressive and it is increasingly difficult to believe that such disparate concerns will ever converge to form a respectable domain of social scientific enquiry. So does the area have any real sociological substance? Where is its focus and what is its future? Are we talking of developing theoretical perspectives and associated methodologies or is this, as some suspect, a case of the emperor's new clothes, with 'the environment' serving as a usefully vague wrapping for work already under way? Here, in the privacy of a postscript, we can face such questions directly.

I have no intention of responding to this challenge by mapping the boundaries of environmental social theory. Such an exercise would be incomplete the moment it was finished. In any case, environmental challenges spread far and wide, presenting opportunities for multidisciplinary enquiry which subject-based definitions cannot encompass. For the time being, therefore, it is tempting to favour limitlessly broad understandings of 'the environment', allowing a thousand flowers to bloom rather than risking any prematurely restrictive definition. This open approach has its advantages but it is important not to dodge the definitional issues entirely. If the 'global environment' is so loose a term that anything can be accommodated under its generous umbrella the questions outlined above will be too slippery to address. One practical solution is to take the chapters in this volume as a point of departure. Drawing upon this material we can begin to construct a considered response to those nagging jibes about the substance, status and future of sociological involvement in global environmental debate.

256

Looking back through the collection, we can see that there is some pattern to the themes addressed. More than that, we can detect the unfolding of two distinct research agendas each preoccupied with characteristically different issues and each propelled by different histories and purposes.

The first is shaped by established sociological interests. The starting point here is the proposition that *the 'global environment' has something to offer social theory*. Recognition of global environmental questions revitalizes well-worn topics and sheds fresh light on the familiar outlines of traditional sociological theory. As suggested in Chapter 1, 'old' conflicts may be 'rendered more tractable' in the process (Benton and Redclift: 7). Ideas relating to nature, to natural resources or to the green movement are thus appropriated by those whose sociological identities have already been formed around other more conventional themes. This is an exciting process and environmental sociology is flourishing where ecological issues mesh with existing interests. Accordingly, niches of environmental research can be found scattered through social theory, in political sociology, in the sociology of science, and so on. In the long run, these pockets of activity may actually change the disciplines of which they are a part. Currently fragmented themes may merge and new 'genuinely' environmental theories develop, but for now the agenda is firmly directed by perceived opportunities for appropriation.

However, this is not the limit of social scientific involvement in the global environmental arena and social theorists are also to be found engaging with issues which initially featured on other people's research agendas rather than simply incorporating them. The sociological imagination is consequently brought to bear upon subjects ranging from acid rain through biodiversity to energy conservation, radio-activity, wildlife preservation and waste disposal. In this more outward-looking context, contributions are made to the sociological understanding of problems which have no part in the conventional repertoire. This type of environmental sociology has a different ethos and is directed towards a different audience. Its programme reflects the concerns and priorities of the outside world rather than those derived from within the discipline itself. The starting point here is the claim that *sociology has something to offer to our understanding of the environment*.

It would be wrong to make too much of this distinction but the notion that sociologists are both *incorporating* and *engaging* helps make sense of their contributions to environmental debate. Such a distinction also allows us to examine the characteristics of emerging environmental agendas, to consider the scope of environmental sociology, and to contemplate its future within the discipline.

INCORPORATING THE GLOBAL ENVIRONMENT

Let us begin by looking at the ways in which social theorists have responded to the broad interdisciplinary challenge presented by global environmental change. Environmental issues now seeping into social science tend to be those of most obvious relevance for current theoretical concerns. Sociologists have consequently latched on to the environment at a number of apparently disconnected points: nature has become a theme for social theorists; green parties are of interest to those researching social movements; environmental consumption becomes an issue for consumer studies; and the 'environmental globe' finds a place for itself in the sociology of globalization. Each point of contact has its own history and its own followers.

Not surprisingly, those involved in the sociological appropriation of the environment have their own implicit understandings of what 'the environment' refers to. The chapters in this book illustrate a range of possible interpretations. Leslie Sklair, for example, defines the realm of the environment with reference to an assembly of eight 'critical issues': water quality, genetic diversity, toxic contamination, and the like. In this approach the environment figures as a series of actual or potential disaster areas. By comparison, Barbara Adam tends to equate the environment with things natural, with natural processes and natural cycles, or with animals, plants and inorganic matter. Here focus is on the interaction between people and nature. Risk and science, alienation from the natural world, and ecological imbalance are important preoccupations all relevant to an understanding of the 'human capacity to have an impact on the environment' and 'loss of control over the effects' (Adam: 104). Alternatively, nature can be seen as an organizing category, one leg of a dualistic opposition inhibiting proper comprehension of the interweaving of social, cultural and ecologial conditions (Benton: 42). This is evidently not the same understanding of nature as that which informs analyses of the distribution and use of natural resources. The projects of exploring the physical and social limits of sustainability or of examining, for instance, the fossil-fuel basis of industrial society (Redclift and Woodgate: 56) belong within different traditions. From such perspectives the important thing about the environment is that it limits and/or enables certain patterns of social life which in turn have implications for the distribution of environmental costs and benefits.

As these few examples illustrate, different writers incorporate very different environments. Here, as elsewhere, the seemingly shared vocabulary of environmental debate is infected through and through by prior interests and approaches. This is neither surprising nor necessarily problematic. What is curious, though, is the degree to which the 'incorporationist' literature then underplays these basic differences, a feature which deserves further attention. Popular appeals to the environment are frequently used in support of quite contradictory propositions: motorways are thus in the environmental interest or not, depending on where we stand. The environmental vocabulary is also immensely flexible, permitting discussion of eco-friendly washing powder whilst also providing the terms for profound analyses of social and political change. But it is strange to encounter social scientists, of all people, talking about whales, motorways, and ozone layers in more or less the same environmental breath. The kinds of things that can be known about sand lizards and global energy crises are of quite a different order, yet the practical consequences of this rather obvious observation are routinely swept away by the totalizing terminology of environmental discourse. What has happened to methodological awareness of the different levels and dimensions of environmental debate, and how is it that sociologists have so readily accepted the apparent unity of environmental issues?

The simplest explanation is that social scientists are addressing different aspects of the same phenomena: in other words there really is some uniting environmental thread. We can argue, for instance, that there is a link between people recycling bottles and global environmental change just as there is a link between our understanding of the natural world and the way in which we conceptualize environmental threat. Studies of culture and ideology are thus as much about global environmental change as are detailed investigations of recycling industries.

In any event, sociologists of the environment have much to gain by cultivating the notion that their subject matter (whatever that may be) is of global significance. 'Global labelling' (Buttel and Taylor: 242–3) is, it seems, as useful for social scientists as it is for environmental movements. Effective mobilization of what Buttel and Taylor call the 'dread factors' depends on linking quite specific areas of social investigation with a vague sense of environmental threat. This linking process is handled in different ways. One pattern, neatly illustrated by Steven Yearley, is to begin by locating the chosen subject within the broad environmental domain (he refers initially to 'the study of the environment'), redefining the meaning of this step by step (from 'environment' to 'the environmental movement', to 'environmental groups' and 'ecological issues'), until it relates to the topic in hand – in this case the Greens and their place within the sociology of social movements. Another tactic is to draw upon a wide range of environmental issues for purposes of illustration. Providing that the underlying argument hangs together, the effect is to create a sense that the many and varied examples do indeed have something in common. Despite their individual differences, the chapters by Adam, Jacobs and Sklair all exhibit some of these tendencies.

While the adoption of global environmental terminology has its benefits, the catch is that it conceals issues of fundamental social significance. This is explicitly recognized by some contributors. Michael Jacobs, for instance, underlines the complexity of environmental valuation, arguing that it requires a 'multiplicity of approaches' (Jacobs: 85), and observing that assessments of the value of Wyoming bighorn sheep cannot be made in the same terms as those used to put prices on air quality or on pleasant country views. Redclift and Benton also acknowledge that 'The "problems" in the environment, and their anticipated "solution" depends on the level at which environmental processes are specified: at locality, regional, national or international level' (Benton and Redclift: 17). This reminds us that not all environmental risks are global. More importantly, it reminds us that the global agenda is imbued with cultural and political interests which we are in danger of taking for granted.

Swayed perhaps by the emotional pull of environmental anxiety, social theorists seem to be losing at least some of their sociological senses. Their common reliance on the grand vocabulary sustains the illusion that all are talking about the same thing but blots out divisions and differences to which they might otherwise attend. In practice, however, all-encompassing environmental vocabularies are used in systematically different ways as environment-related insights are incorporated into already defined areas of social enquiry. Rhetorical acceptance of an undifferentiated package of environmental issues therefore co-exists alongside a series of rather narrow lines of sociological enquiry. This may explain, in part, why sceptics find it so difficult to accept the possibility of environmental sociology: it sounds as if everything should be included, but when it comes to the crunch practitioners are actually pursuing a rag-bag of apparently unrelated themes.

These doubts apart, the environment may yet have a lasting impact on social theory. There are two 'incorporationist' scenarios to consider here. In the first, environmental sociology achieves sub-disciplinary status and an acknowledged, if limited, place within social theory. In the second, the process of incorporating environmental issues has a pervasive influence, reshaping and maybe even shattering the paradigms of the unsuspecting parent discipline.

Exploring the first option, we can already detect moves towards sub-disciplinary legitimation. Given that the sociology of green movements is environmental sociology, environmental sociology may in due course come to be the sociology of green movements (Yearley).

Taking a more historical route, careful sieving of the works of Marx, Weber and Durkheim may reveal ideas and arguments around which to construct a 'new' environmental analysis (Redclift and Woodgate). Nature can be 're-discovered' in this way (Benton) and, in similar fashion, studies of nation-states and transnational systems may prove to be of ecological significance (Sklair). The scene is then set for the emergence of a recognized sub-discipline or, more likely, a series of sub-disciplines linked only by their common reference to the global environment. The prospect of a 'green ghetto' looms.

This first scenario probably underestimates the indigestibility of environmental insights. Can the sociological body simply gobble these up without changing any of its internal preoccupations? Efforts to incorporate the environment may well demand a re-thinking of inherited assumptions (Benton and Redclift: 2) and there is evidence that such changes are under way. Ted Benton and Michael Redclift note, for example, that 'Environmental awareness in the North, however ambiguous its origins, marks a shift in thinking about development' (p. 16). In this context, theories are revised as anxieties about environmental cost overshadow previous concerns about the limits to growth. The side effects of environmental enquiry are even more dramatic in the economic field. What started life as an ordinary exercise in turning 'the environment into a commodity which can be analysed just like other commodities' (Jacobs: 69) has clearly got out of hand, revealing limitations in traditional theoretical frameworks and creating opportunities for previously marginalized points of view. Something similar may happen in sociology if, as Barbara Adam hopes, awareness of the character of contemporary environmental crises 'permeate[s] our social science assumptions, our theories and our methods' (Adam: 108). Ted Benton goes further, first observing that 'in order to do worthwhile sociological research on the "material" dimension of environmental issues, the basic conceptual legacy of the sociological traditions has to be radically re-worked', and then setting out to do just that (Benton: 29). In the process, he begins to transcend conventional categories, establishing a distinctly different way of looking at nature and culture.

Environmental issues may have lasting consequences for those sub-sectors into which they are incorporated but this does not mean that their influence will be felt across the discipline as a whole. In this context, the impact of environmental awareness is contained by the same network of sub-disciplinary boundaries as that which permits its development. Enclaves of environmental enquiry emerge as extensions of established areas of social research and the resulting insights make little difference to those working within other fields. Questions about inter-relationships between pockets of environmental study are consequently bracketed out. Thus far, more attention has been paid to the question 'What does an acknowledgement of the global environment mean for what we thought before?' than to the question 'How do bits of environmental thinking fit together?' This is hardly surprising. To the extent that 'incorporationists' latch on to only those environmental issues which relate to existing preoccupations, their real ambitions have always lain elsewhere.

ENGAGING WITH THE GLOBAL ENVIRONMENT

Sceptical queries about the substance and significance of environmental sociology meet with a different response from those who believe they have a special contribution to make to the understanding of specifically environmental problems. Framed in terms of engagement, rather than incorporation, these questions take on a new meaning.

The first, most striking characteristic of the agenda of engagement is that areas of research are deemed to be important not because of the weight they might carry back home, back, that is, in the heartlands of existing social theory, but because of their perceived relevance for contemporary environmental issues. Prior (though often implicit) assessment of the causes and characteristics of environmental pressure permits a rank ordering of environmental crises and it is this which shapes the direction of social scientific involvement. Such an approach brings with it a whole host of new topics, some of which are evidently more pressing than others. Buttel and Taylor conclude, for instance, that key environmental interactions are best understood in terms of political economy. From their perspective, more political economy is what is required (Buttel and Taylor: 248). In similar vein they outline the terms of a more explicit sociology of environmental science. In making this case they note that sociologists have devoted themselves to the analysis of abstract branches of science 'whose subject matter is perceived to be equally insulated from societal forces' (Buttel and Taylor: 236). This is an appropriate methodological strategy if the purpose is that of examining the social construction of knowledge but it is not the best route if the aim is to understand the political and technological characteristics of environmental science. If the environment is to be centre stage, sociologists will have to engage with different types of science. Social researchers are thus positively encouraged to attend to certain issues and to concentrate their efforts on theoretical and practical enquiries presumed to be of environmental significance.

Equipped with an all-purpose kitbag of sociological tools, 'engagers' then set off into the environmental wilderness, inventing models and constructing analyses as best they can. This can be tricky, of course, for there is no familiar body of sociological literature upon which to build and no guarantee that such engagement will be seen as legitimate sociological activity. In this context the problem is not one of establishing the 'environmental' relevance of research – 'engagers' have less need of 'global labelling' than their 'incorporationist' counterparts – but of justifying sociological exploration of seemingly 'non-sociological' issues. The substantive topics to be tackled are far removed from the usual catalogue of family and gender, work and education, deviance and social inequality, and it is indeed difficult to imagine what a social theory of, say, waste disposal would involve. The challenges of transcending 'dualistic oppositions between subject and object, meaning and cause, mind and matter, human and animal, and, above all, culture (or society) and nature' (Benton: 29) and of addressing the 'material' dimension of environmental issues are equally daunting. Where do we begin with energy crises? What can be said about Chernobyl? In simple terms there are not recognizable sociological issues and our sceptical critics would probably reject the enterprise of engagement on these grounds alone. But sociology is not entirely defined by books already on the shelf. If we see it also as a method, a technique for understanding environmental issues, then we must conclude that 'engagers' are doing real sociological work. More than that, we might even argue that engagement is a precondition for any genuinely environmental sociology.

Before getting too enthusiastic about these refreshingly enterprising and even innovative forms of sociological thought we should reflect, for a moment, on the nature and origin of the environmental agenda with which 'engagers' engage. Taking social theory to the heart of the environmental debate is all very well, but to whose debate are we contributing? What assumptions and interests lie buried within that seemingly self-evident hierarchy of more and less central environmental issues?

Environmental change presents as many challenges for social as for natural sciences but there can be little doubt that natural science, whether cast as the cause of our environmental troubles, as the means by which we know them (Yearley: 162), or the source of their solution, has made the running. In this context, social science has a typically subordinate role. As Brian Wynne observes, 'It either provides information to the natural sciences on human activities which perturb the natural processes, or takes the natural science predictions as given and then works out the social and economic consequences' (Wynne: 170). The danger, then, is that 'engagers' engage (or are engaged) as underlabourers, filling a few well-defined gaps in the natural science agenda and providing the human stuffing for economic and natural scientific models. In studying what policy-makers believe to be the social barriers to environmentally beneficial actions, social scientists confirm a view of the social world as no more than an obstacle to technological progress. Furthermore, they inadvertently fall into the role of tour guides, navigating their natural science colleagues around the intricacies of social life. This is not in itself problematic, but it may become so if it proves to be the limit of the sociological contribution. A second related risk is that social scientists will set aside their own theoretical concerns in the rush to 'get on with the job' of tackling what they agree to be important environmental problems.

Of course, underlabouring is not the only mode of 'engagement'. Wynne, for example, takes natural scientific frameworks as the *subject* of his enquiry, examining in detail the social organization of scientific knowledge and ferreting out the '[n]ormative responsibilities and commitments . . . concealed in the "natural" discourse of the science' (Wynne: 181). Angela Liberatore's chapter provides us with another example of critical engagement. Far from being subsumed within a natural science agenda, she examines the very mechanisms which drive that programme, showing, as it happens, that economists, political scientists, and technologists have a crucial part to play in its formation and funding. Such analyses clearly contribute to the sociology of science and knowledge, though the fact that they do so with specific reference to environmental issues is largely irrelevant from a traditional disciplinary perspective. The problem is that there is no ready-made audience for the environmental insights of sociological enquiry and, having left 'real' sociology behind them, critical 'engagers' must then contend with the routine discomforts of interdisciplinary isolation. Studies of the social organization of particular knowledge systems inevitably meet with a measure of incomprehension from those not sharing the same starting point or the same capacity to see the practical and theoretical significance of assumptions embedded in their everyday scientific practice. For these and other reasons, natural scientists and policy-makers may not be especially receptive to this kind of sociological attention. So where do social theoretical understandings of the construction of environmental science and policy fit in to non-sociological programmes of research and development? It is doubtful whether they can have much direct effect. Social science has already been cast as a small supporting act and efforts to re-write the script are probably doomed to failure.

To summarize, we would have to agree that many of those 'engaging' with environmental issues deal with unfamiliar subject matter. If their work is sociological it is so because sociology is a method and a way of thinking, not because their studies directly relate to any recognizable body of sociological knowledge. In straying so far from the traditional path, then, social theorists may indeed lose their

264

way, finding themselves rejected by their more conventional colleagues and enslaved or marginalized by the environmental agenda with which they seek to engage.

SUSTAINING DEVELOPMENTS

Where does this discussion leave us? Is environmental sociology just a loose assembly of assorted issues or is it here to stay?

Much depends on how we view the nature of the enterprise: do we think it involves applying sociological ways of thinking to a whole range of environment-related issues or is it a matter of searching out environmental themes which can be subsumed within sociology? Although I have used 'incorporationism' and 'engagement' as contrasting categories these are not exclusive positions. As the chapters in this book illustrate, opening paragraphs asking about sociology's role in environmental debate are swifly followed by questions about the challenges which the global environment presents for social theory. The two strands are intertwined within the collection and within its chapters. The routes of 'engagement' and 'incorporation' do, however, lead us in different directions and they do have different implications for the future of environmental sociology.

What is the stuff of environmental sociology, does it qualify as sociology and what is environmental about it? 'Incorporationist' literature has well-established sociological credentials but it is often difficult to believe that its themes are really environmental. The sheer range of issues which now attract the environmental label is in itself sufficient to generate suspicion. If the environment can be all these different things then how can we take it seriously? This question is doubly difficult to respond to when the process of incorporation creates isolated pockets of environmental sociology and, at the same time, fragments the subject as an aspiring field of enquiry in its own right. Disconnected areas of 'environmental' thought then develop around specific environmental issues as these are framed by specific sub-disciplinary preoccupations. In this context we should probably admit that there is no such thing as 'environmental sociology', while still making a strong case that social theorists have much to learn from careful consideration of environmental issues. By contrast, 'engagers' subscribe to an agenda which has ready-made environmental credentials. True, the list of environmental themes is still very broad but 'engagers' at least can defend their involvement with reference to other non-sociological assessments of environmental relevance. 'Engagers' talk to strangers from other disciplines and deal in unfamiliar topics: is what they do really sociological? The answer is surely 'yes', with the implication that there can be a sociology of the environment but only if we view sociology as a distinctive way of thinking about phenomena and only if we accept the terms of the agenda with which particular engagers engage.

What, then, of the future? Those 'incorporating' the global environment would appear to have rather brighter immediate prospects than those 'engaging' with it. After all, 'incorporationists' can look forward to a new sub-discipline and even anticipate 'far-reaching changes at the very heart of sociology' (Adam: 93). By comparison, 'engagers' are much more likely to be shunted to the margins of their own discipline and left to talk to scientists and policy-makers about things which these groups don't really want to hear. While their chances of single-handedly dislodging technological determinism are slight, they too might aspire to broadening the base of social theory. Providing that 'engagers' can hang on to their critical capacities, retaining their conceptual integrity within the environmental fray, they may be able to legitimize the sociological study of entirely new issues thus changing the agenda of social science as well as modifying the perceptions and choices of non-sociological actors. That certainly would be a sustainable development in environmental sociology.

INDEX